Watershed Environmental Degradation and Sustainable Development

Tongxin Zhu

Watershed Environmental Degradation and Sustainable Development

The Case of the Yellow River Basin

Springer

Tongxin Zhu
Department of Environment, Sustainability and Geography
University of Minnesota
Duluth, MN, USA

ISBN 978-3-032-02447-3 ISBN 978-3-032-02448-0 (eBook)
https://doi.org/10.1007/978-3-032-02448-0

This Springer imprint is published by the registered company Springer Nature Switzerland AG
The registered company address is: Gewerbestrasse 11, 6330 Cham, Switzerland

Preface

A watershed, or a river basin, governs the flow of water, nutrients, and sediments within its boundaries. These flow processes extend through the channel networks, creating cascading impacts from upstream to downstream areas. A large river basin is organized as a hierarchical structure, composing multiple-level sub-watersheds. There are similarities and differences in landscape and ecological functions and services across these levels. All the large river basins in the world have transitioned from a status solely acted by natural factors in the pre-human era to a stage characterized by intimate human–nature interactions nowadays. Human activities, such as clearing natural vegetation for agricultural production, exert impacts on the watershed functions and processes that a variety of environmental problems may arise. Sustainable development in a watershed aims to mitigate these environmental problems and to achieve environmental, social and economic sustainability while enhancing people's livelihoods.

The Yellow River is the second longest river in China and the sixth longest river in the world. The Yellow River Basin is a cradle of Chinese civilization. Currently, more than 100 million people reside in the watershed. Land use changes driven by rapid population growth in recent history have led to the significant watershed-wide degradation of the sensitive and fragile ecosystems and caused a variety of environmental problems, such as soil erosion, desertification, grassland degradation, wetland losses, salinization, water shortage and pollution, which have adversely affected the economic and social development. In the past few decades, particularly since the late 1990s, large-scale and intensive environmental remediation programs have been carried out in the basin to tackle some of the worst environmental problems in the world.

Tremendous amounts of research have been conducted in the Yellow River Basin by experts from a wide range of fields, such as agriculture, biology, environment, geology, geography, economics, engineering, and hydrology, among others. These studies resulted in a plethora of publications. This book aims to provide a comprehensive review of research on environmental degradation and management strategies geared toward sustainable development in the Yellow River Basin. The objectives of this book are as follows:

First, this book provides a comprehensive review of the existing studies on the environment issues of the Yellow River Basin, aiming at synthesizing key research findings and highlighting areas that require further exploration. It can serve as a valuable reference for researchers and practitioners in future studies of the region's environmental issues.

Second, this book explores the basin's complex environmental challenges across multiple scales, shaped by its unique climate, soil types, topography, and human activities. It examines the interconnected nature of various environmental issues and their cascading impacts throughout the watershed.

Third, this book evaluates the extensive environmental restoration and sustainable development efforts in the Yellow River Basin, which may offer valuable experience to guide practices in other river basins around the world.

Duluth, MN, USA Tongxin Zhu

Competing Interests The author has no competing interests to declare that are relevant to the content of this manuscript.

Contents

Chapter 1
Introduction

Contents

1.1 Human Impacts on the Environment

No place on this planet has not been impacted by human activities directly or indirectly. Humans have altered the natural landscape by clearing vegetation for agriculture, constructing infrastructure, extracting mineral resources, and burning fossil fuels, among other activities. These activities have led to a range of environmental problems, including loss of natural vegetation and biodiversity, soil erosion, water contamination, desertification, air pollution, and global warming. With the rapid growth in human population, the exploitation and consumption of resources like food, energy, and water have significantly increased (Macklin & Lewin, 2015). The most significant impacts on the environment results from direct land use (Thomas Jr., 1956; Lambin et al., 2003). Land use, on the one hand, fulfills essential human needs like food and resources, but on the other hand, it can cause environmental degradation (Foley et al., 2005). Over the last three centuries, approximately 1200 million hectares of forests and woodlands have been cleared (Richards, 1990; Ramankutty & Foley, 1999); steppes, savannas, and grasslands have shrunk from around 3200 million to 1800–2700 million hectares (Lambin et al., 2003). Meanwhile, global cropland area has increased from 265 million hectares in 1700 to 1471 million hectares by 1990. During the same period, pastureland expanded more than sixfold, growing from 524 million to 3451 million hectares (Goldewijk,

T. Zhu, *Watershed Environmental Degradation and Sustainable Development*,
https://doi.org/10.1007/978-3-032-02448-0_1

2001). In the US, since the European settlement, more than 50% of the original wetlands have been converted to agricultural lands (Mitsch & Gosselink, 2015).

Land use changes, especially the clearing of natural vegetation for agricultural production, lead to habitat degradation, fragmentation, and edge effects. These transformations contribute to biodiversity loss and accelerate species extinction (Lindenmeyer & Burgman, 2005). It is estimated that approximately nine million species of plants, animals, protists, and fungi inhabit the Earth (Cardinale et al., 2012a, 2012b). However, the global rate of species extinction over the past 50 years is at least tens to hundreds of times higher than the average rate over the past ten million years (IPBES, 2019). A report compiled by the Intergovernmental Science-Policy Platform on Biodiversity and Ecosystem Services (IPBES) of the United Nations Environment Programme (UNEP), which draws on evidence collected from roughly 15,000 scientific and government studies over the past five decades, found that 25 percent of mammals, more than 40 percent of amphibian species, nearly 33 percent of sharks, and 25 percent of plant groups are threatened with extinction.

Without human disturbance, natural erosion rates are generally balanced by the rate of soil formation over the long term. However, changes in land use have accelerated soil loss, particularly in arable lands. This acceleration has intensified sheet, rill, and gully erosion on hillslopes, as well as increased mass movements—such as landslides, earthfalls, and debris flows—on valley slopes in the watersheds. Agricultural land in semi-arid and arid regions is also prone to wind erosion. Studies from around the globe show that erosion rates from agricultural fields are on average 1–2 orders of magnitude higher than the rates of soil production, erosion under natural vegetation, and long-term geological erosion rates (Montgomery, 2007). For instance, in Brazil's savanna environment, the removal of natural vegetation resulted in an increase in soil loss from 0.1 Mg/ha per year to 12.4 Mg/ha per year (Oliveira et al., 2015). Global soil loss from arable land due to anthropogenic erosion is estimated to range from 20 Pg to over 100 Pg per year (GSP, 2017; Borrelli et al., 2017; Panagos et al., 2020). Additionally, agricultural activities lead to soil compaction from repeated cultivation, which reduces infiltration and increases surface runoff. This can lead to an increased risk of flooding. According to Knox (1993), agricultural land use in the Upper Mississippi River Basin has increased flood magnitudes by one to two orders of magnitude compared to pre-agricultural, natural conditions.

Soil erosion causes nutrient depletion, reduces topsoil depth, and contributes to land degradation. The eroded sediments often carry pesticides and excess nutrients—primarily from fertilizers—leading to the pollution of aquatic ecosystems. It is estimated that 80% of water pollution in the United States comes from non-point sources, primarily agricultural lands, while only 20% originates from point sources (USDA, 1989).

A significant portion of eroded sediments is deposited along the way before reaching their final destination, such as the ocean. It's estimated that the worldwide transfer of sediment to the oceans from continents via rivers spans from 3.9 to 19.1 billion tons per year, according to studies by Holeman (1968), Hatono and Yoshimura (2020), and Milliman and Farnsworth (2011). This significant variation in estimates arises from the lack of sufficient sediment measurement facilities, the assumptions of the models used, and the fluctuating amounts of sediment transported by rivers over time.

Soil erosion rates in river basins worldwide far exceed the rates of sediment deposition in the oceans. This disparity is primarily due to the sediment delivery ratios (SDR) of drainage basins (Walling, 1983, 1999). SDRs are affected not only by the physical characteristics of a basin, including its size and slope, but also by human-made structures within the basin, such as dams constructed across river channels. Worldwide, approximately 40,000 large dams and 800,000 smaller ones have significantly reduced the sediment flow to coasts by an estimated 1.4 ± 0.3 billion tons annually (Syvitski et al., 2005; Keiser et al., 2005). More than five hundred hydropower dams have been constructed in the Amazon River basin alone, causing hydrophysical and biotic disturbances that impact the basin's floodplains, estuary and sediment plume (Latrubesse et al., 2017).

Urbanization has rapidly increased worldwide, with more than 55% of the global population now residing in cities, a trend expected to continue (UN DESA, 2018). Satellite mapping reveals urban sprawl covers 0.3 to 2.4% of the Earth's terrestrial surface (Potere & Schneider, 2007). Urban area increases 625,000 to 1,039,000 km^2 from 2000 to 2012. Impervious surfaces in urban areas disrupt hydrological cycles, alter stormwater flow and seasonal patterns (Band et al., 2005; Hung et al., 2018), raise temperatures, and contribute to climate change (Kalnay & Cai, 2003; Macklin & Lewin, 2015). Urbanization also causes land subsidence due to groundwater withdrawals (Abidin et al., 2015), degrades soil health and biodiversity (Yang & Zhang, 2015), and contribute to both water and air pollution (Hobbie et al., 2017; Seinfeld, 1989; Cohen et al., 2004).

Fossil fuel burning is a significant source of greenhouse gas emissions, exacerbating global warming (Lashof & Ahuja, 1990). Global warming causes sea levels to rise (Douglas, 1991; Church & White, 2006; Church et al., 2013; Cazenave et al., 2014). Without intervention, sea levels could rise by 0.52 to 0.98 meters by the end of the century (IPCC, 2013), potentially inundating low-lying coastal areas. Global temperature rises will further melt permafrost (Koven et al., 2011; Chadburn et al., 2017), change precipitation patterns (Huang et al., 2013; Lehmann et al., 2015), intensify hurricanes (Pielke Jr et al., 2005; Mousavi et al., 2011), contribute to desertification (Ma et al., 2021), and reduce biodiversity (Malcolm et al., 2006; Cardinale et al., 2012a, 2012b).

1.2 Watershed Processes

A watershed is a natural topographic unit that drains water into a common outlet, such as an ocean or a lake. A large watershed, also known as a drainage basin, is a nested hierarchical system comprising multiple levels of subwatersheds. Water, soil, and vegetation interact with each other across various spatial scales within a watershed through physical, chemical, and biological processes that provide the ecological functions and services critical for sustaining the life of all species within its boundaries.

Water movement, one of the key processes in a watershed, transports sediment, circulates vital nutrients for plant growth, and shapes the watershed‘s landscape. Flow generation exhibits great complexity and heterogeneity in different locations and across various scales. Water exchanges occur in situ between the atmosphere and land surfaces through precipitation, evaporation, and transpiration. Water moves through the soil via infiltration and percolation to reach groundwater. On a hillslope, water moves laterally as overland flow (including infiltration-excess overland flow and saturation overland flow) on the surface, interflow between the ground surface and water table, and baseflow below the water table. Infiltration-excess flow, also known as Hortonian overland flow (Horton, 1945), predominates in semi-arid and arid areas. The heterogeneity of Hortonian overland flow generation is largely controlled by spatial variations in rainfall and infiltrability, which are affected by soil conditions, patches of vegetation, slope gradients, and other factors. Conversely, saturation overland flow, also known as Dunne's overland flow, commonly occurs in humid environments and is controlled by the dynamic expansion of the saturated area, or partial contributing area (Dunne and Black, 1970a, 1970b). Surface and subsurface flow from hillslopes enter channels, route through stream networks, and eventually exit at the watershed outlet. Hydrological behavior across different watershed levels shows both scale invariance and scale dependence (Gupta & Waymire, 1998).

Erosion is a major process in a watershed. Natural erosion rates, also referred to as geological erosion rates, are usually low but vary due to different climatic and geomorphic conditions (Vanwalleghem et al., 2006). For instance, Nearing et al. (2017) compiled data from studies conducted in six different areas of the US and indicated that geological erosion rates ranged from 0.035 to 1.9 Mg ha^{-1} $yr.^{-1}$. Dunne et al. (1978) estimated that the natural erosion rate averaged 0.0084 mm/yr., equivalent to a yield of 0.22 Mg ha^{-1} $yr.^{-1}$ of sediments and solutes in semi-arid Kenya throughout most of the Tertiary period. Hewawasam et al. (2003) estimated that geological erosion rates in the tropical highlands of Sri Lanka range from 0.13 to 0.3 Mg ha^{-1} $yr.^{-1}$. Eroded materials can be transported by water flows, wind, and gravity, and deposited on hillslopes, within gullies, streams and river channels, and in water bodies, forming depositional landforms such as alluvial plains and river deltas.

Nutrients are vital for plant growth. Nutrient cycling and movement within a watershed involve different paths and processes at different scales. Natural processes of nutrients input to a watershed include weathering processes from bedrocks, wet and dry deposition from the atmosphere, biological processes through photosynthesis and N fixation (Attiwill & Adams, 1993). Plants absorb nutrients through uptake from soils and through fixation from the atmosphere and return nutrients back to soils as litter through decomposition. Nutrient cycling is a critical function of an ecosystem, but at a widely varying rate under different plant communities, with a decreasing order of grass < deciduous forest < evergreen forest. Any place within a watershed where the input of a nutrient(s) exceeds the output of a nutrient(s) over a period is a nutrient sink, such as wetland, estuary, and delta. Nutrients are transported by water and sediment within and out of a watershed. Fluvial processes redistribute nutrients on the hillslope, move nutrients from the upper watershed to the lower watershed via river networks, and also bilaterally exchange nutrients between river channels and the riparian landscapes (Peterjohn & Correll, 1984). During the movement of nutrients, there are multiple exchanges between sediments and water, and between flows and their adjacent aquatic and terrestrial environments (Vought et al., 1994).

Plant communities vary within a watershed and are affected by different factors at different scales. At the macroscale, climate is the most important factor in determining the types of plant biomes present (e.g. forest, grassland, desert) (Woodward et al., 2004). At the local scale, factors such as soil, hydrology, and topography interact to shape the distribution, composition, and functioning of biological communities. The growth of plant communities, in turn, affects soil formation, erosion, and water movement (Meyer et al., 1992; Sebastiá, 2004; Hou et al., 2016).

1.3 Sustainable Development and Watershed Management

Watershed management involves the effective practice of sustainable development and sustainable management is the ultimate goal of watershed management. The concept of sustainability has a long and evolving history, shaped by scholars across diverse disciplines—such as environmental science, economics, social sciences, and political science—and has been extensively examined in the literature (Kidd, 1992; Scoones, 2007; Giovannoni & Fabietti, 2013; Portney, 2015). The most widely accepted definition of sustainability is the principle of fulfilling current needs without jeopardizing the ability of future generations to meet theirs (World Commission on Environment and Development, 1987). Sustainability is supported by three core pillars: environmental, economic, and social sustainability (United Nations, 2002). Environmental sustainability focuses on restoring and protecting natural ecosystems and resources to ensure their availability for current and future generations. Some of the key practices are reducing soil erosion, improving water quality, conserving vegetation and biodiversity, and promoting the use of renewable energy sources. Economic sustainability seeks to promote long-term economic growth through the efficient use of resources while ensuring environmental protection. This can be achieved through a range of strategies, including the development of sustainable agriculture and industry, improvement of water use efficiency, promotion of economic diversification, transition from unsuitable land use to sustainable land management, and the provision of technological training for local communities. Social sustainability strives to reduce inequality, enhance education, and promote social responsibility and environmental awareness. Achieving sustainability across the three pillars calls for a systems thinking approach to understand the intricate interconnections between environmental, social, and economic systems. The United Nations proposed 17 Sustainable Development Goals (SDG) in

2015 as part of the 2030 Agenda for Sustainable Development. While the SDGs are presented as distinct objectives, many are interrelated - advancement in one area can support progress in others (Mensah, 2019). Increasingly, people have emphasized the importance of adopting an integrated approach to sustainability (Giovannoni & Fabietti, 2013).

Watershed management is also an evolving concept that has developed over millennia—from its early focus on managing water resources to a more holistic approach that encompasses the management of the entire watershed ecosystem, including land, water, biodiversity, and other resources, for ecological, social, and economic objectives (Ffolliott et al., 2002; Wang et al., 2016).Watershed management and sustainability are closely interconnected, as effective watershed management supports long-term environmental, social, and economic goals. It is aligned with key United Nations Sustainable Development Goals (SDGs), including SDG 6—access to safe water and sanitation—and SDG 15—'Life on Land,' which emphasizes sustainable land use, forest management, and biodiversity conservation.

Watershed management has its own distinctive features. The nature of watershed connectivity, where energy, matter, and organisms interact through various pathways, underscores the importance of integrated management practices. This connectivity necessitates coordinated efforts to address disturbances in any part of the watershed, as these can have downstream effects. For example, to mitigate sediment accumulation in the Yellow River's lower reaches, conservation measures are needed in the Loess Plateau, its sediment source area. Likewise, addressing the Gulf of Mexico's dead zones requires controlling nutrient runoff from agricultural lands in the Mississippi watershed. A watershed approach stands out as an effective strategy for managing environmental challenges due to the interconnected processes within its domain.

A large river basin is a nested system composed of multi-scale sub-watersheds. Watershed management involves both scaling up and scaling down strategies, each of which is essential and complementary for achieving sustainable outcomes. Scaling up connects local efforts to regional or national frameworks, while scaling down tailors broader strategies to specific local conditions. Managing a large river basin often requires large-scale, government-initiated projects that address critical environmental challenges. These key initiatives are typically led at the national level to ensure coordination, resource allocation, and long-term impact. A well-known example of large-scale watershed management led by a national government is the Tennessee Valley Authority (TVA) in the United States. Initiated in the 1930s, the TVA was designed to address severe environmental degradation, flooding, deforestation, and poverty in the Tennessee River watershed. Through a series of integrated infrastructure projects—such as dams, reservoirs, and reforestation efforts—the TVA successfully improved water management, reduced soil erosion, generated hydroelectric power, and supported regional economic development. This case illustrates how government-led initiatives can play a pivotal role in addressing complex environmental and socio-economic challenges across a large watershed (Hargrove, 1994). Small watersheds are the building blocks of larger river basins.

Effective management at the small scale helps maintain the overall stability and functionality of the system as activities in a small watershed can influencing water quality, sediment load, and ecosystem health downstream. It also enables the adaptation of management strategies to specific local environmental conditions (Geng et al., 2010; Shi et al., 2012). In turn, the successful small watershed management strategies can be replicated at larger scales.

Effective watershed management requires a holistic, integrated approach. This includes cross-sectoral coordination across land use, water resources utilization, environmental remediation and protection, and social-economic development. It emphasizes upstream–downstream collaboration, the remediation of related environmental issues, and the active involvement of local communities, government agencies, and other stakeholders. It needs to strike a balance between short-term needs and long-term priorities by aligning economic development with environmental sustainability. The Yellow River basin has faced a range of environmental challenges—such as soil erosion, flooding, drought, land degradation, wetland loss, declining vegetation cover, and climate change—that hinder sustainable development (Fu, 1989; Gao, 2016; Li et al., 2019; Mostern, 2021; Omer et al., 2020; Pietz, 2015). In recent decades, environmental remediation and ecological restoration efforts have been carried out to address those issues. The Yellow River basin serves as a great example of both environmental degradation and the mitigation efforts undertaken to promote sustainable development at the scale of a large river basin.

A vast amount of research has been conducted in the Yellow River basin across a wide range of fields, such as agriculture, environment, engineering, soils, ecology, geography, geology, biology, hydrology, among others. This book aims to offer a comprehensive analysis of environmental issues and sustainable development within the basin by reviewing existing research and identifying gaps for future studies. and offering recommendations for watershed management.

This book is structured into four main sections. First, it will have an overview of the physical environmental conditions of the Yellow River basin. Second, it will examine the paleo-environments and their evolutions in the upper, middle, and lower reaches of the basin. Understanding paleo-environments is important for evaluating the respective roles of natural and anthropogenic impacts on contemporary environments and predicting the effect of climate changes on the future environment. Third, it will examine the major environmental issues in the basin, their underlying causes, and the strategies used for mitigation and remediation, along with their effectiveness. Fourth, the book will examine the intertwined relationships among different environmental problems and the impacts from upstream to downstream, as well as discuss a few large policy-driven projects to address these interrelated problems.

References

Abidin, H. Z., Andreas, H., Gumilar, I., & Wibowo, I. R. R. (2015). On correlation between urban development, land subsidence and flooding phenomena in Jakarta. *Proceedings of the International Association of Hydrological Sciences, 370*, 15–20.

Attiwill, P. M., & Adams, M. A. (1993). Nutrient cycling in forests. *New phytologist, 124*(4), 561–582.

Band, L. E., Cadenasso, M. L., Grimmond, C. S., Grove, J. M., & Pickett, S. T. (2005). Heterogeneity in urban ecosystems: patterns and process. In *Ecosystem functions in heterogeneous landscapes* (pp. 257–278). Springer.

Borrelli, P., Robinson, D. A., Fleischer, L. R., et al. (2017). An assessment of the global impact of 21st century land use changes on soil erosion. *Nature Communications, 8*, 2013.

Cardinale, B., Duffy, J., Gonzalez, A., et al. (2012a). Biodiversity loss and its impact on humanity. *Nature, 486*, 59–67.

Cardinale, B. J., Duffy, J. E., Gonzalez, A., Hooper, D. U., Perrings, C., Venail, P., Narwani, A., Mace, G. M., Tilman, D., Wardle, D. A., & Kinzig, A. P. (2012b). Biodiversity loss and its impact on humanity. *Nature, 486*(7401), 59–67.

Cazenave, A., Dieng, H. B., Meyssignac, B., Von Schuckmann, K., Decharme, B., & Berthier, E. (2014). The rate of sea-level rise. *Nature Climate Change, 4*(5), 358–361.

Chadburn, S. E., Burke, E. J., Cox, P. M., Friedlingstein, P., Hugelius, G., & Westermann, S. (2017). An observation-based constraint on permafrost loss as a function of global warming. *Nature Climate Change, 7*(5), 340–344.

Church, J. A., & White, N. J. (2006). A 20th century acceleration in global sea-level rise. *Geophysical Research Letters, 33*, L01602.

Church, J. A., Clark, P. U., Cazenave, A., Gregory, J. M., Jevrejeva, S., Levermann, A., Merrifield, M. A., Milne, G. A., Nerem, R. S., Nunn, P. D., & Payne, A. J. (2013). *Sea level change*. PM Cambridge University Press.

Cohen, A. J., Anderson, H. R., Ostro, B., Pandey, K. D., Krzyzanowski, M., Künzli, N., Gutschmidt, K., Pope, C. A., III, Romieu, I., Samet, J. M., & Smith, K. R. (2004). Urban air pollution. In *Comparative quantification of health risks: global and regional burden of disease attributable to selected major risk factors* (Vol. 2, pp. 1353–1433). WHO.

Douglas, B. C. (1991). Global sea level rise. *Journal of Geophysical Research: Oceans, 96*(C4), 6981–6992.

Dunne, T., & Black, R. D. (1970a). Partial area contributions to storm runoff in a small New England watershed. *Water Resources Research, 6*(5), 1296–1311.

Dunne, T., & Black, R. D. (1970b). An experimental investigation of runoff production in permeable soils. *Water Resources Research, 6*(2), 478–490.

Dunne, T., Dietrich, W. E., & Brunengo, M. J. (1978). Recent and past erosion rates in semi-arid Kenya. *Zeitschrift für Geomorphologie, 29*, 130–140.

Ffolliott, P. F., Baker, M. B., Edminster, C. B., Dillon, M. C., & Kora, K. L. (Eds.). (2002). *Land stewardship through watershed management, perspective for 21st Century* (p. 137). Kluwer Academic/Plenum Publishers.

Foley, J. A., DeFries, R., Asner, G. P., Barford, C., Bonan, G., Carpenter, S. R., Chapin, F. S., Coe, M. T., Daily, G. C., Gibbs, H. K., & Helkowski, J. H. (2005). Global consequences of land use. *Science, 309*(5734), 570–574.

Fu, B. (1989). Soil erosion and its control in the Loess Plateau of China. *Soil Use and Management, 5*(2), 76–82.

Gao, J. (2016). Wetland and its degradation in the Yellow River Source Zone. In *Landscape and ecosystem diversity, dynamics and management in the Yellow River Source Zone* (pp. 209–232). Springer.

Geng, Y., Mitchell, B., Tsuyoshi, F., & Nakayama, T. (2010). Perspectives on small watershed management in China: the case of Biliu. *International Journal of Sustainable Development & World Ecology, 17*(2), 172–179.

Giovannoni, E., & Fabietti, G. (2013). What is sustainability? A review of the concept and its applications. In *Integrated reporting: Concepts and cases that redefine corporate accountability* (pp. 21–40). Springer.

Goldewijk, K. K. (2001). Estimating global land use change over the past 300 years: the HYDE database. *Global Biogeochemical Cycles, 15*(2), 417–434.

GSP. (2017). *Global soil partnership endorses guidelines on sustainable soil management*. GSP.

Gupta, V. K., & Waymire, E. (1998). Spatial variability and scale invariance in hydrologic regionalization. In *Scale dependence and scale invariance in hydrology* (pp. 88–135). Cambridge University Press.

Hargrove, E. C. (1994). *Prisoners of Myth: The Leadership of the Tennessee Valley Authority, 1933–1990*. Princeton University Press.

Hatono, M., & Yoshimura, K. (2020). Development of a global sediment dynamics model. *Progress in Earth and Planetary Science, 7*(1), 1–15.

Hewawasam, T., von Blanckenburg, F., Schaller, M., & Kubik, P. (2003). Increase of human over natural erosion rates in tropical highlands constrained by cosmogenic nuclides. *Geology, 31*(7), 597–600.

Hobbie, S. E., Finlay, J. C., Janke, B. D., Nidzgorski, D. A., Millet, D. B., & Baker, L. A. (2017). Contrasting nitrogen and phosphorus budgets in urban watersheds and implications for managing urban water pollution. *Proceedings of the National Academy of Sciences, 114*(16), 4177–4182.

Holeman, J. N. (1968). The sediment yield of major rivers of the world. *Water Resources Research, 4*, 737–747.

Horton, R. E. (1945). Erosional development of streams and their drainage basins; hydrophysical approach to quantitative morphology. *Geological Society of America Bulletin, 56*(3), 275–370.

Hou, J., Wang, H., Fu, B., Zhu, L., Wang, Y., & Li, Z. (2016). Effects of plant diversity on soil erosion for different vegetation patterns. *Catena, 147*, 632–637.

Huang, P., Xie, S. P., Hu, K., Huang, G., & Huang, R. (2013). Patterns of the seasonal response of tropical rainfall to global warming. *Nature Geoscience, 6*(5), 357–361.

Hung, C. L. J., James, L. A., & Carbone, G. J. (2018). Impacts of urbanization on stormflow magnitudes in small catchments in the Sandhills of South Carolina, USA. *Anthropocene, 23*, 17–28.

IPBES, 2019. Summary for policymakers of the global assessment report on biodiversity and ecosystem services of the Intergovernmental Science-Policy Platform on Biodiversity and Ecosystem Services.

IPCC. (2013). Annex III: Glossary [Planton, S. (Ed.)]. In T. F. Stocker, D. Qin, G.-K. Plattner, M. Tignor, S. K. Allen, J. Boschung, A. Nauels, Y. Xia, V. Bex, & P. M. Midgley (Eds.), *Climate change 2013: The physical science basis. Contribution of Working Group I to the Fifth Assessment Report of the Intergovernmental Panel on Climate Change*. Cambridge University Press.

Kalnay, E., & Cai, M. (2003). Impact of urbanization and land-use change on climate. *Nature, 423*(6939), 528–531.

Keiser, J., De Castro, M. C., Maltese, M. F., Bos, R., Tanner, M., Singer, B. H., & Utzinger, J. (2005). Effect of irrigation and large dams on the burden of malaria on a global and regional scale. *The American Journal of Tropical Medicine and Hygiene, 72*(4), 392–406.

Kidd, C. V. (1992). The evolution of sustainability. *Journal of Agricultural and Environmental Ethics, 5*, 1–26.

Knox, J. C. (1993). Large increases in flood magnitude in response to modest changes in climate. *Nature, 361*(6411), 430–432.

Koven, C. D., Ringeval, B., Friedlingstein, P., Ciais, P., Cadule, P., Khvorostyanov, D., Krinner, G., & Tarnocai, C. (2011). Permafrost carbon-climate feedbacks accelerate global warming. *Proceedings of the National Academy of Sciences, 108*(36), 14769–14774.

Lambin, E. F., Geist, H. J., & Lepers, E. (2003). Dynamics of land-use and land-cover change in tropical regions. *Annual Review of Environment and Resources, 28*(1), 205–241.

Lashof, D. A., & Ahuja, D. R. (1990). Relative contributions of greenhouse gas emissions to global warming. *Nature, 344*(6266), 529–531.
Latrubesse, E. M., Arima, E. Y., Dunne, T., Park, E., Baker, V. R., d'Horta, F. M., Wight, C., Wittmann, F., Zuanon, J., Baker, P. A., & Ribas, C. C. (2017). Damming the rivers of the Amazon basin. *Nature, 546*(7658), 363–369.
Lehmann, J., Coumou, D., & Frieler, K. (2015). Increased record-breaking precipitation events under global warming. *Climatic Change, 132*, 501–515.
Li, Y., Xie, Z., Qin, Y., & Zheng, Z. (2019). Responses of the Yellow River basin vegetation: climate change. *International Journal of Climate Change Strategies and Management, 11*(4), 483–498.
Lindenmeyer, D., & Burgman, M. (2005). *Practical conservation biology*. CSIRO Publishing. ISBN 0-643-09089-4.
Ma, X., Zhu, J., Yan, W., & Zhao, C. (2021). Projections of desertification trends in Central Asia under global warming scenarios. *Science of the Total Environment, 781*, 146777.
Macklin, M. G., & Lewin, J. (2015). The rivers of civilization. *Quaternary Science Reviews, 114*, 228–244.
Malcolm, J. R., Liu, C., Neilson, R. P., Hansen, L., & Hannah, L. E. E. (2006). Global warming and extinctions of endemic species from biodiversity hotspots. *Conservation Biology, 20*(2), 538–548.
Mensah, J. (2019). Sustainable development: Meaning, history, principles, pillars, and implications for human action: Literature review. *Cogent Social Sciences, 5*(1), 1653531.
Meyer, S. E., García-Moya, E., & Lagunes-Espinoza, L. D. C. (1992). Topographic and soil surface effects on gypsophile plant community patterns in central Mexico. *Journal of Vegetation Science, 3*(3), 429–438.
Milliman, J. D., & Farnsworth, K. L. (2011). *River discharge to the coastal ocean: A global synthesis*. Cambridge University Press.
Mitsch, W. J., & Gosselink, J. G. (2015). *Wetlands*. John Wiley & Sons.
Montgomery, D. R. (2007). Soil erosion and agricultural sustainability. *Proceedings of the National Academy of Sciences, 104*(33), 13268–13272.
Mostern, R. (2021). *The Yellow River: A natural and unnatural history*. Yale University Press.
Mousavi, M. E., Irish, J. L., Frey, A. E., Olivera, F., & Edge, B. L. (2011). Global warming and hurricanes: the potential impact of hurricane intensification and sea level rise on coastal flooding. *Climatic Change, 104*, 575–597.
Nearing, M. A., Xie, Y., Liu, B., & Ye, Y. (2017). Natural and anthropogenic rates of soil erosion. *International Soil and Water Conservation Research, 5*(2), 77–84.
Oliveira, P. T. S., Nearing, M. A., & Wendland, E. (2015). Orders of magnitude increase in soil erosion associated with land use change from native to cultivated vegetation in a Brazilian savannah environment. *Earth Surface Processes and Landforms, 40*(11), 1524–1532.
Omer, A., Zhuguo, M., Zheng, Z., & Saleem, F. (2020). Natural and anthropogenic influences on the recent droughts in Yellow River Basin, China. *Science of the Total Environment, 704*, 135428.
Panagos, P., Borrelli, P., & Robinson, D. (2020). FAO calls for actions to reduce global soil erosion. *Mitigation and Adaptation Strategies for Global Change, 25*, 789–790.
Peterjohn, W. T., & Correll, D. L. (1984). Nutrient dynamics in an agricultural watershed: observations on the role of a riparian forest. *Ecology, 65*(5), 1466–1475.
Pielke, R. A., Jr., Landsea, C., Mayfield, M., Layer, J., & Pasch, R. (2005). Hurricanes and global warming. *Bulletin of the American Meteorological Society, 86*(11), 1571–1576.
Pietz, D. A. (2015). *The Yellow River: The problem of water in modern China*. Harvard University Press.
Portney, K. E. (2015). *Sustainability*. MIT Press.
Potere, D., & Schneider, A. (2007). A critical look at representations of urban areas in global maps. *GeoJournal, 69*(1), 55–80.
Ramankutty, N., & Foley, J. A. (1999). Estimating historical changes in global land cover: croplands from 1700 to 1992. *Global Biogeochemical Cycles, 13*(4), 997–1027.

Richards, J. F. (1990). Land transformation. In B. L. Turner et al. (Eds.), *The earth as transformed by human action* (pp. 163–178). Cambridge Univ. Press.
Scoones, I. (2007). Sustainability. *Development in Practice, 17*(4–5), 589–596.
Sebastiá, M. T. (2004). Role of topography and soils in grassland structuring at the landscape and community scales. *Basic and Applied Ecology, 5*(4), 331–346.
Seinfeld, J. H. (1989). Urban air pollution: state of the science. *Science, 243*(4892), 745–752.
Shi, Z. H., Ai, L., Fang, N. F., & Zhu, H. D. (2012). Modeling the impacts of integrated small watershed management on soil erosion and sediment delivery: a case study in the Three Gorges Area, China. *Journal of Hydrology, 438*, 156–167.
Syvitski, J. P., Vorosmarty, C. J., Kettner, A. J., & Green, P. (2005). Impact of humans on the flux of terrestrial sediment to the global coastal ocean. *Science, 308*(5720), 376–380.
Thomas, W. L., Jr. (1956). *Man's Role in Changing the Face of the Earth* (p. 1193). University of Chicago Press.
United Nations. (2002). *Report of the World Summit on Sustainable Development, Johannesburg, South Africa, 26 August–4 September 2002*. United Nations.
United Nations Department of Economic and Social Affairs (UN DESA). (2018). *2018 revision of world urbanization prospects*.
USDA. (1989). *The second RCA appraisal*. USDA.
Vanwalleghem, T., Bork, H. R., Poesen, J., Dotterweich, M., Schmidtchen, G., Deckers, J., Scheers, S., & Martens, M. (2006). Prehistoric and Roman gullying in the European oess belt: a case study from central Belgium. *The Holocene, 16*(3), 393–401.
Vought, L. B. M., Dahl, J., Pedersen, C. L., & Lacoursiere, J. O. (1994). Nutrient retention in riparian ecotones. *Ambio, 23*, 342–348.
Walling, D. E. (1983). The sediment delivery problem. *Journal of Hydrology, 65*(1–3), 209–237.
Walling, D. E. (1999). Linking land use, erosion and sediment yields in river basins. In *Man and river systems: The functioning of river systems at the basin scale* (pp. 223–240). Springer.
Wang, G., Mang, S., Cai, H., Liu, S., Zhang, Z., Wang, L., & Innes, J. L. (2016). Integrated watershed management: evolution, development and emerging trends. *Journal of Forestry Research, 27*, 967–994.
Woodward, F. I., Lomas, M. R., & Kelly, C. K. (2004). Global climate and the distribution of plant biomes. *Philosophical Transactions of the Royal Society of London. Series B: Biological Sciences, 359*(1450), 1465–1476.
World Commission on Environment and Development. (1987). *Our Common Future*. Oxford University Press.
Yang, J. L., & Zhang, G. L. (2015). Formation, characteristics and eco-environmental implications of urban soils–A review. *Soil Science and Plant Nutrition, 61*(sup1), 30–46.

Chapter 2
The Yellow River Basin

Contents

2.1 Introduction

The Yellow River is the second longest river in China and the sixth longest in the world. The river's current name was adopted in the Western Han Dynasty (202 B.C. to 8 A.D.) in reference to its heavily sediment-laden waters. Prior to this period, it was widely referred to as "Da He," or the Great River, according to Ren and Zhu (1994). The China Exploration and Research Society identifies the origin of the Yellow River at the coordinates 34°29′31″N 96°20′25″E in the Bayan Har Mountain on the Qinghai-Tibet Plateau. The Yellow River stretches 5464 km and traverses seven provinces and two autonomous regions, flowing into the Bohai Sea, a semi-enclosed segment of the NW Pacific Ocean in northern China (Fig. 2.1). It is widely acknowledged that the dawn of early Chinese civilization took root in the Yellow River basin during the Neolithic era, bestowing upon it the title "the cradle of Chinese Civilization" (Lawler, 2009). Because the river's prone to flood in the lower reaches, causing widespread destruction, has led to its moniker "China's Sorrow" (Clapp, 1922).

T. Zhu, *Watershed Environmental Degradation and Sustainable Development*,
https://doi.org/10.1007/978-3-032-02448-0_2

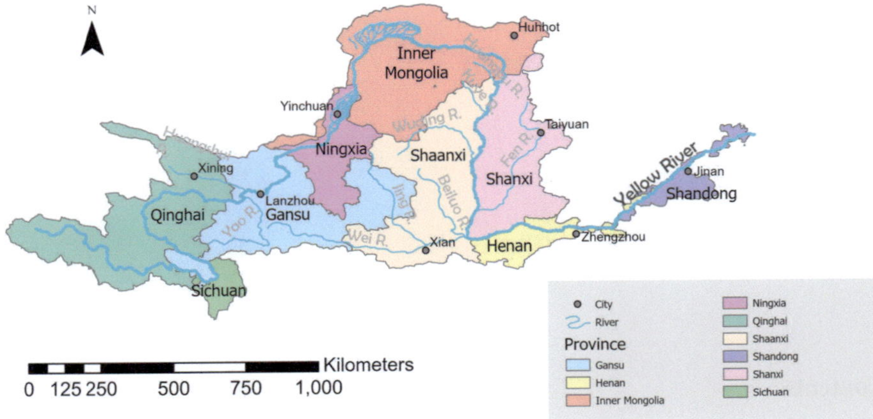

Fig. 2.1 The Yellow River Basin

As the world's tenth-largest river basin, the Yellow River Basin covers a drainage area of 752,000 km^2, according to Wu et al. (2004). The basin is home to a population of 114 million people (YRCC, 2013). The Yellow River Basin, recognized for its fragile environmental conditions and extensive human utilization of natural resources, faces significant environmental challenges such as soil erosion, vegetation destruction, flooding and land degradation (Wohlfart et al., 2016).

2.2 The Formation of the Yellow River

The rise of the Tibetan Plateau has significant impacts on the formation of the Yellow River and the climates of the Yellow River basin (Chang et al., 1989; Ding et al., 2002; Ding et al., 2022; Kusky et al., 2007; Qiu, 2008; Royden et al., 2008). The Tibetan Plateau, also known as the Qinghai-Tibet Plateau,is encircled by mountains on every side, forming an intermontane plateau. It is flanked to the south by the Himalayas, to the north by the Kunlun Mountains, to the east by the Hengduan Mountains, and to the west by the Karakoram Mountains. Renowned as the world's largest and highest plateau, the Tibetan Plateau spans approximately 2,600,000 square kilometers, with an average altitude of around 4500 meters above sea level, earning it the nickname "Roof of the World."

The formation and evolutionary processes of the Yellow River in the Tibet Plateau are still subjects of debate and ongoing research (Lan et al., 2021). Craddock et al. (2010) believe that river erosion began in the early Pleistocene, around 1.8 million years ago. They suggest that the erosional process was triggered by the growth of nearby lake systems, leading to overflow events that integrated the modern Yellow River network. In contrast, Liu (2020) proposes an earlier commencement of the river's erosion, dating back to the late Miocene. Nie et al. (2015) contend that the formation of the river's upper section occurred during the Pliocene, while Lin et al. (2001) asserts its development took place in the Eocene. The Yellow River has carved deep gorges, over 600 meters deep, along the Tibet Plateau‘s edge, reflecting the plateau's uplift and the concurrent deepening of the river's gorges in its upper stretches. Han et al. (2013) identified three terraces within the river's source area, formed mainly from late Middle Pleistocene to Holocene, demonstrating complex erosional and depositional histories.

The middle segment of the Yellow River is characterized by its distinctive rectangular course around the Ordos Plateau. This area, known as the Ordos block, is encircled by the Yinchuan graben to the west, the Hetao graben to the north, and the Shanxi graben to the southeast. By the late Miocene-Pliocene period, the Ordos Block underwent a significant geological transition from a basin to a plateau (Yue et al., 2007). Geological records indicate that the formation of the Yellow River's angular bend around the Ordos block probably occurred during the late Miocene to early Pliocene era. This was primarily due to processes of folding and uplift, which are believed to be closely associated with the ongoing collision and penetration of the Indian subcontinent into Eurasia, along with rifting processes adjacent to the Ordos block (Lin et al., 2001). Prior to the formation of the rectangular course, the Yellow River flowed eastward through the Weihe River.

The lower reaches of the Yellow River extend into the North China Plain, which is a vast rift basin that was formed through downfaulting. The Sanmenxia Gorge serves as a natural conduit, linking the lower section of the Yellow River with its upper and middle sections. The Sanmenxia Gorge was formed by westward capturing the paleolake and the gorge may have been formed by lake overflow during the period 3.63–1.4 Ma (Hu et al., 2017). Before this linkage was established, sediment deposits in the North China Plain primarily originated from local sources. Following its integration, sediments from the upper and middle reaches of the Yellow River have been transported downstream, contributing to the sedimentation in the lower reaches and the eastern China Sea.

The timeline for the full integration of the Yellow River continues to be a contentious topic within the scientific community (Zhang et al., 2021;Chen et al., 2022), with proposed dates ranging from the Miocene period (Lin et al., 2001), to the Middle Pleistocene (Hu et al., 2016; Li et al., 1996; Xiao et al., 2020; Zhu, 1989), or the Late Pleistocene (Zhang et al., 2018).

2.3 Topographic Features

The Yellow River Basin has an average elevation of 1887 meters, with its terrain ranging from sea level to over 6000 meters (Fig. 2.2). This area features a distinct three-tiered elevation structure, starting from the river's source and extending to its mouth (Fig. 2.3). It is divided into three main sections: a high-altitude plateau at the source, a mid-elevation plateau, and a low-lying alluvial plain. The river's slope

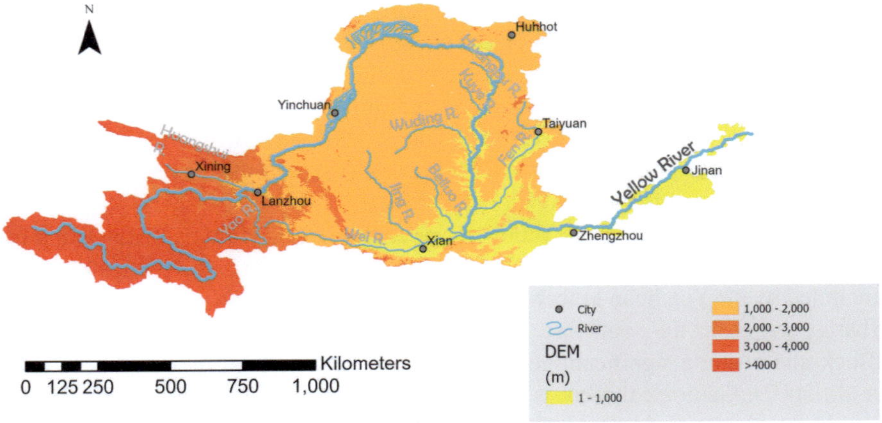

Fig. 2.2 DEM in the Yellow River Basin

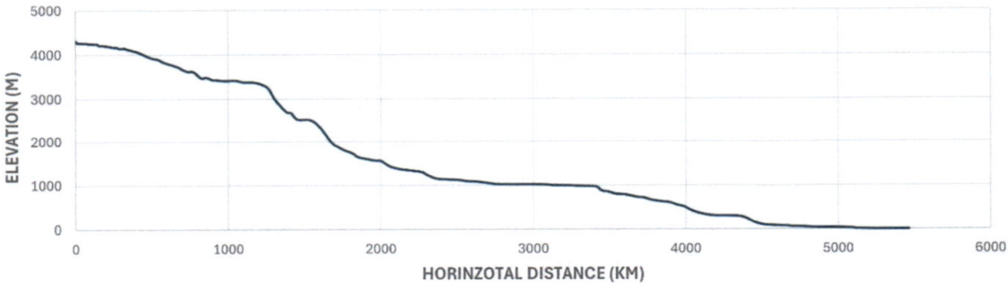

Fig. 2.3 Longitudinal profile of the Yellow River

varies significantly across its sections, with gradients of 0.96%, 0.095%, and 0.0099% for the upper, middle, and lower reaches, respectively (Yu, 2006).

The terrain of the source region, located in the Tibetan Plateau, is marked by snow-capped mountains, alpine plateaus, lakes, and deep gorges. This high-altitude plateau, primarily within Qinghai Province, has an average elevation of 3870 meters. Its slopes gradually descend northeastward until they reach the plateau's edge, where the elevation sharply drops due to deep ravines created by tectonic activity and fluvial erosion. Stretching from the river's source to Lanzhou, the elevation decreases from 4600 meters to 1236 meters over a distance of 3472 km (Cheng et al., 2012).

The Loess Plateau dominates the middle section of the Yellow River Basin, known for its severe erosion, which has carved a landscape filled with gullies. This region, covering Gansu, Ningxia, Inner Mongolia, Shaanxi, and Shanxi Provinces, has an average elevation of around 1480 meters. The plateau features distinct landforms: *yuan* (elevated flatlands), *liang* (long, gully-filled hills), and *mao* (rounded hills with steep sides). In this section, the Yellow River descends from 1236 meters to 79 meters over a distance of 1206 km (Cheng et al., 2012). Additionally, the region includes plains adjacent to the river and its tributaries, such as the Yinchuan

Plain to the west, the Hetao Plain to the north, and the Guanzhong Plain to the south. High mountains also border the Loess Plateau on its western, eastern, and southern sides.

The lower section of the Yellow River Basin is part of the alluvial North China Plain, where the river has historically shifted its course frequently due to flooding. In this stretch, the river descends from 79 meters to sea level over a distance of 786 km. Due to heavy siltation, a "hanging river" has formed in some parts of the lower channel, with the riverbed sitting as much as 10 meters higher than the surrounding plain.

The Yellow River delta is formed by the sediment carried from the Yellow River basin, particularly from the middle reaches of the Loess Plateau. Due to the frequent breaches and shifts of the lower river channels, the delta has experienced numerous avulsions and is composed of several superlobes, each consisting of several smaller lobes created by shifts in the distributary channels (Xue, 1993). The expansion of the delta area over time is directly related to the runoff and sediment discharge of the Yellow River, which is affected by the anthropogenic and natural factors, such as climate variability, soil erosion and conservation, and dam construction in the basin.

2.4 Climates

According to the Köppen climate classification, the major types of climate in the Yellow River basin are Alpine Tundra (EH) and monsoon-influenced subarctic (Dwc) in the source region; cold arid desert (BWk), cold semi-arid steppe (BSk), monsoon-influenced warm-summer humid continental (Dwb), and hot-summer humid continental (Dwa) in the upper and middle reaches; and monsoon-influenced humid subtropical (Cwa) and monsoon-influenced temperate humid continental (Dwa) in the lower reaches (Beck et al., 2018).

The source region's climate is characterized by extended, frigid winters and brief, temperate summers, accompanied by dry winds, abundant sunshine, and significant diurnal temperature variations. The average annual temperature spans from 0 to 8 °C, with a gradient of increasing cold from the southeast to the northeast. January stands as the coldest month, with average temperatures below −10 °C, whereas July is the warmest, with averages between 10–20 °C. Annual precipitation is about 300 mm, which tends to increase from the northeast towards the southeast, with most of the annual precipitation occurring during summer. Rainfall intensity is typically low (Hu et al., 2012; Li et al., 2016). The period from October to April is marked by subfreezing temperatures, leading to the formation of widespread permafrost across 85% of the source region (Jin et al., 2022).

The middle reaches lie the Loess Plateau, known for its cold winters and warm to hot summers. January's average monthly temperatures vary from less than −6 °C in the western and northwestern parts to 0 °C in the southeast, while July's averages fluctuate from 15 °C in the northwest to 27 °C in the southeast. Annual precipitation increases from approximately 200 mm in the northwest to over 700 mm in the southeast, with 60–70% of annual precipitation occurring from June to September. Summer storms are characterized by high intensity, often leading to significant erosion. Spring and winter are predominantly dry and windy, with a limited amount of snowfall. The region's potential evaporation exceeds its annual precipitation by as much as 1100 mm (Zhou et al., 2020).

The lower reaches of the Yellow River are influenced by the East Asian Monsoon, with dry, cold winters and hot, wet summers (Li et al., 2019a, 2019b). The average annual temperature is around 15 °C, with January averaging 0.5 °C and July reaching 27.5 °C. Annual precipitation ranges from 600 mm to 700 mm, with the highest rainfall occurring in July and the lowest in December.

2.5 Hydrology and Water Resources

The Yellow River is known for its relatively low flow discharge, influenced by the semi-arid and arid climates, with an average annual discharge of about 58 billion m^3—roughly 1/16 of the Yangtze River's total flow. The annual flow discharge shows an overall decreases in the past 50 years due to climate changes and human activities (Ran & Lu, 2012; Wu et al., 2015; Xu et al., 2002; Yu et al., 2013). Seasonal flow variation has also declined significantly during this time, largely due to the construction of large reservoirs along the Yellow River and its tributaries, which store water during the rainy season and release it during the dry season (Xu et al., 2022). Total water uses in the Yellow River basin reached 41.9 billion m^3 in 2000, representing 80% of the available water resources, far exceeding the international safety threshold of 40% (Guo et al., 2022; Jia et al., 2006). Much of this water usage, over 80%, is dedicated to agricultural irrigation.

Hydrologic measurements by the Yellow River Commission indicate that flow discharge fluctuates along the river from the source region to the lower reaches, rather than increasing consistently with drainage area (Fig. 2.4), due to the combined impacts of water abstraction and reservoir storage.

The source region, located upstream of the Tang Nai Hai (TNH) hydrometric station and covering about 15.4% of the total basin area, contributes approximately 35% of the Yellow River's flow (Hu et al., 2011; Li et al., 2016; Zheng et al., 2007). Snowmelt accounts for up to 40% of the total runoff in spring or about 13% of the river's annual flow (Cuo et al., 2013; Lan et al., 1999). However, research indicates a declining trend in flow discharge in the source region in recent years, largely driven by climate change (Cao et al., 2006; Cuo et al., 2013; Hu et al., 2011).

The upper and middle sections of the Yellow River basin, located downstream of Lanzhou and largely encompassing the Loess Plateau, are characterized by semi-arid and arid climates. These regions have relatively limited water resources but experience high agricultural demand (Wei et al., 2016). The average annual flow discharge at the Longmen station was even lower than that at the Lanzhou hydrologic station, even though the drainage area increased from 23.8% to 65.2% of the total Yellow River basin (Fig. 2.4).

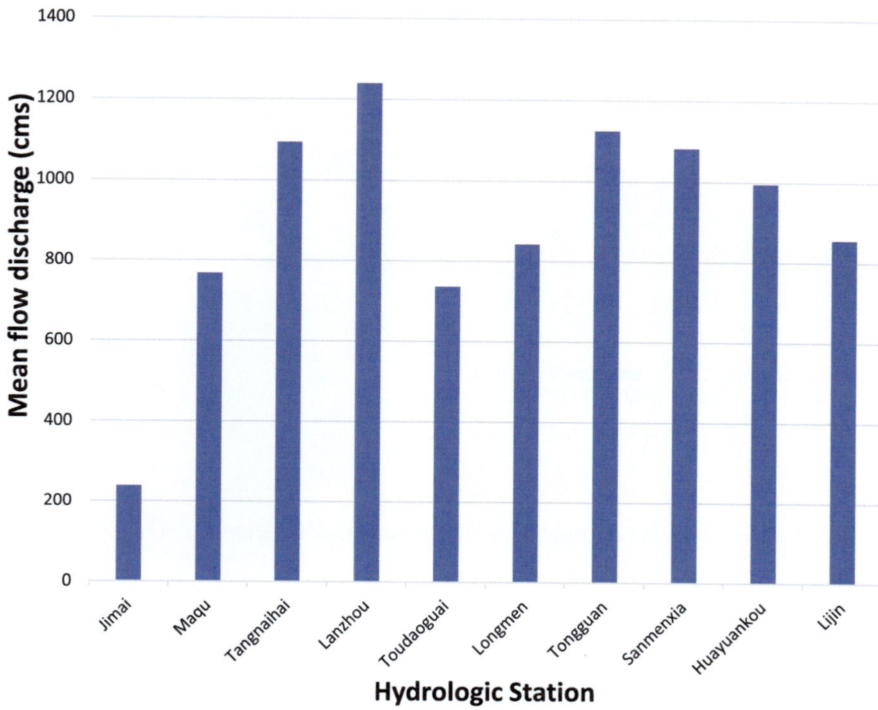

Fig. 2.4 Flow discharge along the Yellow River changes from its source through to its lower reaches over the period of 1952 to 2019

Downstream from the Sanmenxia station, flower discharge further decreases due to water storage in reservoirs and consumption by various sectors, including industry, agriculture, and urban areas. Notably, the river experienced periods of no flow, particularly from 1990 to 2000, with 1997 marking a record 226 days without flow (Cong et al., 2009).

2.6 Land Cover

Figure 2.5 displays the Yellow River basin's land cover, based on satellite imagery data from the International Geosphere-Biosphere Program's Data and Information System, a collaborative project of the U.S. Geological Survey and the European Commission's Joint Research Centre (Loveland & Belward, 1997). This data categorizes the area into 17 distinct classes at a spatial resolution of 1 km (Table 2.1). The major land uses within the basin are grassland, cropland, and shrubland, which constitute 46.1%, 24.9%, and 18.8% of the area, respectively (Fig. 2.6).

In the source region, the major land uses are grassland (75%), barren area (11%), forest (7.4%), and wetland (4.3%) (Liu et al., 2021). Recent studies suggest that

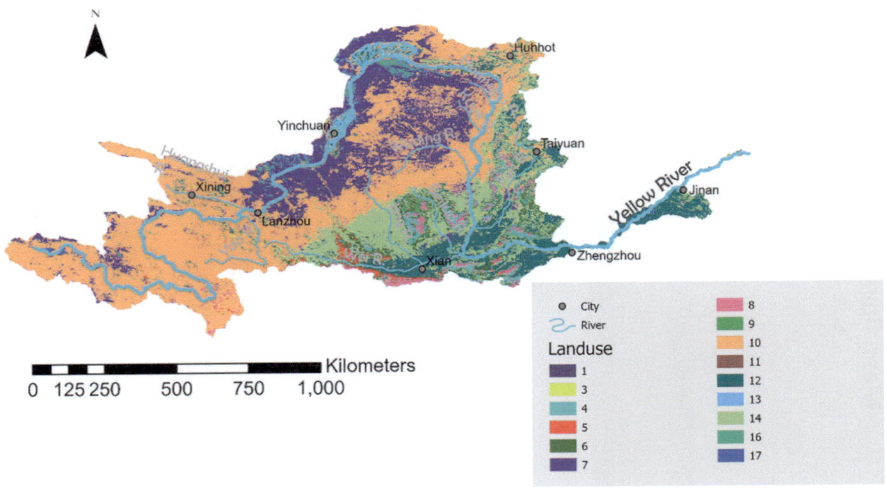

Fig. 2.5 Land use in the Yellow River basin. The classification codes are described in Table 2.1

over the past 50 years, grasslands have deteriorated and desertification has worsened due to climate change (rising temperatures and decreasing precipitation) and human activities, such as overgrazing (Iqbal et al., 2022; Na et al., 2013; Song et al., 2009). The region's vegetation mainly consists of high-altitude grasslands and alpine meadows. High-cold grasslands are primarily composed of species like *Stipa purpurea*, *Achnatherum splendens*, *Stipa breviflora*, *Carex moorcroftii*, and *Littledalea racemosa*, while alpine meadows are dominated by *Kobresia pygmaea*, *Kobresia humilis*, and *Kobresia tibetica* (Wang et al., 2004). Sparse forests also occur in the area, including needleleaf forests on cold mountain slopes with species such as *Picea crassifolia* and *Sabina przewalskii*, along with temperate broadleaf deciduous forests featuring *Populus davidiana*. Shrublands are also present, dominated by subalpine broadleaf evergreen and deciduous shrubs like *Salix oritrepha* and *Caragana jubata* (Wang et al., 2020).

In the upper and middle reaches below the source region, the dominant land uses are cropland (38.8%), grassland (34.5%), and a combination of forests and shrubs (19.7%) (Liu et al., 2021). The main crops in the region are wheat, corn, millet, potatoes, and soybeans. Since the late 1990s, land use has shifted, with a reduction in cropland and an increase in forest and grassland areas, largely due to the implementation of the "Grain for Green" program (Deng et al., 2014; Li et al., 2019a; Li et al., 2019b; Wu et al., 2019; Xiao, 2014). The middle reaches, mainly the Loess Plateau, encompasses diverse vegetation zones—ranging from forest, forest-steppe, and steppe to semi-desert and desert—driven by a gradient of decreasing precipitation and temperature from southeast to northwest (Zhang et al., 2006). The region's forests are primarily secondary, with common species including *Armeniaca sibirica*, *Salix matsudana*, *Populus cathayana*, *Sophora japonica*, *Pinus tabuliformis*, and *Quercus liaotungensis* (Du et al., 2011; Liu et al., 2018). The main shrub species are

Table 2.1 IGBP land cover classification system

Code Land cover categories
1. Evergreen needleleaf forests Lands dominated by needleleaf woody vegetation with a percent cover >60% and height exceeding 2 m. Almost all trees remain green all year. Canopy is never without green foliage
2. Evergreen broadleaf forests Lands dominated by broadleaf woody vegetation with a percent cover >60% and height exceeding 2 m. Almost all trees and shrubs remain green year round. Canopy is never without green foliage
3. Deciduous needleleaf forests Lands dominated by woody vegetation with a percent cover >60% and height exceeding 2 m. Consists of seasonal needleleaf tree communities with an annual cycle of leaf-on and leaf-off periods
4. Deciduous broadleaf forests Lands dominated by woody vegetation with a percent cover >60% and height exceeding 2 m. Consists of broadleaf tree communities with an annual cycle of leaf-on and leaf-off periods
5. Mixed forests Lands dominated by trees with a percent cover >60% and height exceeding 2 m. Consists of tree communities with interspersed mixtures or mosaics of the other four forest types. None of the forest types exceeds 60% of landscape
6. Closed shrublands Lands with woody vegetation less than 2 m tall and with shrub canopy cover >60%. The shrub foliage can be either evergreen or deciduous
7. Open shrublands Lands with woody vegetation less than 2 m tall and with shrub canopy cover between 10% and 60%. The shrub foliage can be either evergreen or deciduous
8. Woody savannas Lands with herbaceous and other understory systems, and with forest canopy cover between 30% and 60%. The forest cover height exceeds 2 m
9. Savannas Lands with herbaceous and other understory systems, and with forest canopy cover between 10% and 30%. The forest cover height exceeds 2 m
10. Grasslands Lands with herbaceous types of cover. Tree and shrub cover is less than 10%
11. Permanent wetlands Lands with a permanent mixture of water and herbaceous or woody vegetation. The vegetation can be present either in salt, brackish, or fresh water
12. Croplands Lands covered with temporary crops followed by harvest and a bare soil period (e.g., single and multiple cropping systems). Note that perennial woody crops will be classified as the appropriate forest or shrub land cover type
13. Urban and built-up lands Land covered by buildings and other man-made structures
14. Cropland/natural vegetation mosaics Lands with a mosaic of croplands, forests, shrubland, and grasslands in which no one component comprises more than 60% of the landscape
15. Snow and ice Lands under snow/ice cover throughout the year
16. Barren Lands with exposed soil, sand, rocks, or snow and never have more than 10% vegetated cover during any time of the year
17. Water bodies Oceans, seas, lakes, reservoirs, and rivers. Can be either fresh or salt water body

Robinia pseudoacacia and *Hippophae rhamnoides*, while grasslands are primarily composed of species like *Medicago sativa L.*, *Setaria viridis*, *Stipa bungeana*, and *Bothriochloa ischaemum* (Xiao et al., 2022; Zhang et al., 2015; Zhang et al., 2016).

In the lower reaches, the main land uses are cropland (68.4%), urban areas (15.7%), and grassland (5.4%) (Liu et al., 2021). The major crops include wheat, corn, cotton, soybeans, and peanuts. The non-cultivated vegetation is mainly found in the river delta wetland and the small riparian wetlands located between the riverbanks. These wetlands support a rich variety of plant species, with more than 500

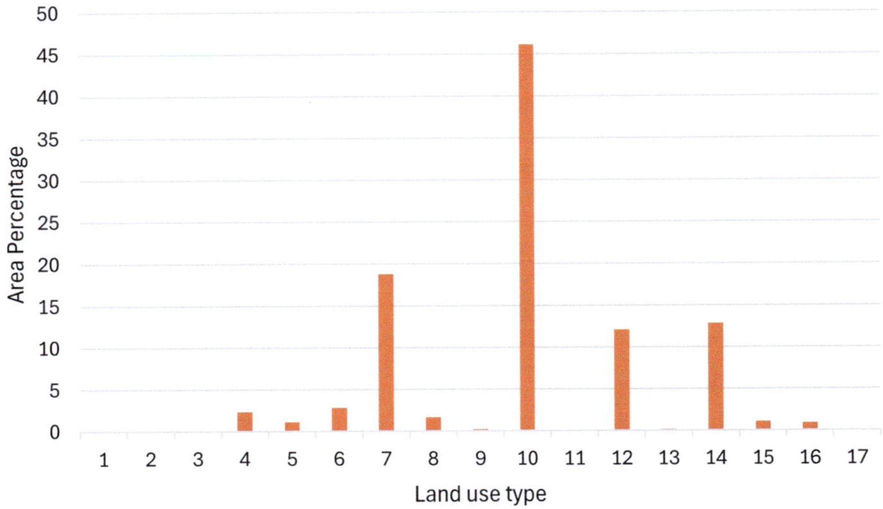

Fig. 2.6 Areal percentage of different landuse in the Yellow River Basin

identified, including dominant species such as *Phragmites australis*, *Tamarix chinensis*, and *Phragmites sage* (Jiao et al., 2014; Wang et al., 2012).

2.7 Surface Materials and Soils

The soils in the Yellow River's source region are primarily composed of alpine meadow and alpine steppe soils, with pockets of sandy and swamp soils (Wang et al., 2007). Since grasslands are the dominant land cover, the Tibetan Plateau's grasslands are considered one of the world's most important carbon reserves (Wang et al., 2002; Yuan et al., 2022). The average concentrations of soil organic carbon (SOC), total nitrogen (TN), and total phosphorus (TP) in these grasslands are 56.71, 4.07, and 0.62 g/kg, respectively. However, these values decrease significantly as grassland degradation worsens in the region (Wang et al., 2016).

The middle reaches of the Yellow River are largely covered by wind-blown deposits (Liu, 1965; Zhang et al., 2020). The loess in this region is classified into sandy loess, loess, and clayey loess zones, from northwest to southeast. The soil has a loamy texture, predominantly composed of silt particles (0.05–0.01 mm), which make up 40% in the north and up to 58% in the south. Fine sand content also increases from north to south, with the highest found in the sandy loam of the north (ranging from 40.8% to 64.9%), while clay particle content increases towards the south up to 20% (Yang et al., 1999). The soil is alkaline with a high $CaCO_3$ content, typically between 6–18% (Zhao et al., 2017). Soil organic matter (SOM), total nitrogen (N), phosphorus (P), and potassium (K) levels vary greatly depending on land use. Vegetation land generally has higher nutrient levels than croplands, and revegetation efforts in recent years enhance these nutrients (Gong et al., 2021).

In the lower reaches of the Yellow River, the terrain is flat and primarily composed of fluvial deposits. According to the FAO-UNESCO classification system, the soil in this region is classified as Calcaric Fluvisols (FAO, 1988). It has a loamy texture, slightly alkaline, and exhibits soil salinization in some areas. Soil organic matter (SOM) contents range from 3.25 to 21.73 g/kg, total nitrogen (N) from 0.22 to 1.28 g/kg, and total phosphorus (P) from 14.64 to 22.98 g/kg (Guo et al., 2017).

2.8 Summary

The Yellow River Basin exhibits significant and complex variability in its environment. Topographically, the river originates in the Qinghai-Tibet Plateau, meanders through the Loess Plateau in its middle section, and ends in a flat floodplain. The climatic conditions along its course vary greatly, starting with alpine climates in its source and upper regions, shifting to semi-arid and arid conditions in the middle, and concluding with a warm, semi-humid climate in the lower basin. The soil composition also changes dramatically from alpine meadow and steppe soils in the source region, to loamy soils in the middle sections derived from loess, and to Fluvisols in the lower reaches. Land use patterns follow suit, with the upper reaches of grasslands and alpine meadows, the middle reaches of a mix of cultivated land and grassland, and the lower reaches of cropland. Regarding water resources, over 35% of the water originates from the source region. However, water availability becomes increasingly strained in the middle and lower reaches due to consumption for agriculture, industry, domestic use, and ecological needs. Overall, the Yellow River Basin faces significant environmental stress, driven by fragile natural conditions and intense human activities. Despite these challenges, the basin supports a significant portion of the country's population and agricultural output.

References

Beck, H. E., Zimmermann, N. E., McVicar, T. R., Vergopolan, N., Berg, A., & Wood, E. F. (2018). Present and future Köppen-Geiger climate classification maps at 1-km resolution. *Nature Scientific Data, 5*, 180214.

Cao, J., Qin, D., Kang, E., & Li, Y. (2006). River discharge changes in the Qinghai-Tibet Plateau. *Chinese Science Bulletin, 51*, 594–600.

Chang, C. F., Pan, Y. S., & Sun, Y. Y. (1989). The tectonic evolution of Qinghai-Tibet Plateau: A review. In *Tectonic Evolution of the Tethyan Region* (pp. 415–476). Springer.

Chen, Q., Liu, X., Zhao, G., Jia, J., Ye, W., Lü, B., Meadows, M., & Guan, J. (2022). 0.2 Ma or 1.2 Ma? Timing of the Linking of the Middle and Lower Reaches of the Yellow River Inferred From Loess-Palaeosol Sequences. *Geophysical Research Letters, 49*(6), e2021GL097510.

Cheng, Q., Han, J., Zhu, D. G., He, C. G., Meng, X. G., Shao, Z. G., Wang, J., & Yu, J. (2012). An analysis of geomorphologic characteristics of the Yellow River source region based on ASTER-GDEM. *Geology in China, 39*(5), 1247–1260.

Clapp, F. G. (1922). The Hwang Ho, Yellow River. *Geographical Review, 12*(1), 1–18.

Cong, Z., Yang, D., Gao, B., Yang, H., & Hu, H. (2009). Hydrological trend analysis in the Yellow River basin using a distributed hydrological model. *Water Resources Research, 45*(7), 1–13.

Craddock, W. H., Kirby, E., Harkins, N. W., Zhang, H., Shi, X., & Liu, J. (2010). Rapid fluvial incision along the Yellow River during headward basin integration. *Nature Geoscience, 3*(3), 209–213.

Cuo, L., Zhang, Y., Gao, Y., Hao, Z., & Cairang, L. (2013). The impacts of climate change and land cover/use transition on the hydrology in the upper Yellow River Basin, China. *Journal of Hydrology, 502*, 37–52.

Deng, L., Shangguan, Z. P., & Sweeney, S. (2014). "Grain for Green" driven land use change and carbon sequestration on the Loess Plateau, China. *Scientific Reports, 4*(1), 7039.

Ding, Z. L., Derbyshire, E., Yang, S. L., Yu, Z. W., Xiong, S. F., & Liu, T. S. (2002). Stacked 2.6-Ma grain size record from the Chinese loess based on five sections and correlation with the deep-sea δ18O record. *Paleoceanography, 17*(3), 5–1.

Ding, L., Kapp, P., Cai, F., Garzione, C. N., Xiong, Z., Wang, H., & Wang, C. (2022). Timing and mechanisms of Tibetan Plateau uplift. *Nature Reviews Earth & Environment, 3*, 1–16.

Du, S., Wang, Y. L., Kume, T., Zhang, J. G., Otsuki, K., Yamanaka, N., & Liu, G. B. (2011). Sapflow characteristics and climatic responses in three forest species in the semiarid Loess Plateau region of China. *Agricultural and Forest Meteorology, 151*(1), 1–10.

FAO. (1988). Soil map of the world: revised legend. In *World soil resources report* (Vol. 60). FAO.

Gong, C., Tan, Q., Liu, G., & Xu, M. (2021). Impacts of species mixture on soil nitrogen stocks in the Loess Plateau of China. *Forest Ecology and Management, 491*, 119145.

Guo, L., Sun, Z., Ouyang, Z., Han, D., & Li, F. (2017). A comparison of soil quality evaluation methods for Fluvisol along the lower Yellow River. *Catena, 152*, 135–143.

Guo, L., Zhu, W., Wei, J., & Wang, L. (2022). Water demand forecasting and countermeasures across the Yellow River basin: Analysis from the perspective of water resources carrying capacity. *Journal of Hydrology: Regional Studies, 42*, 101148.

Han, J. E., Shao, Z. G., Zhu, D. G., Meng, X. G., Jia, Y. U., Wang, J., Lv, R. P., Qiang, C., & He, C. G. (2013). Characteristics of river terraces and formation of the Yellow River in the source region of Yellow River. *Geology in China, 40*(5), 1531–1541.

Hu, Y., Maskey, S., Uhlenbrook, S., & Zhao, H. (2011). Streamflow trends and climate linkages in the source region of the Yellow River, China. *Hydrological Processes, 25*(22), 3399–3411.

Hu, Y., Maskey, S., & Uhlenbrook, S. (2012). Trends in temperature and rainfall extremes in the Yellow River source region, China. *Climatic Change, 110*(1–2), 403–429.

Hu, Z., Pan, B., Guo, L., Vandenberghe, J., Liu, X., Wang, J., Fan, Y., Mao, J., Gao, H., & Hu, X. (2016). Rapid fluvial incision and headward erosion by the Yellow River along the Jinshaan gorge during the past 1.2 Ma as a result of tectonic extension. *Quaternary Science Reviews, 133*, 1–14.

Hu, Z., Pan, B., Bridgland, D., Vandenberghe, J., Guo, L., Fan, Y., & Westaway, R. (2017). The linking of the upper-middle and lower reaches of the Yellow River as a result of fluvial entrenchment. *Quaternary Science Reviews, 166*, 324–338.

Iqbal, M., Wen, J., Masood, M., Masood, M. U., & Adnan, M. (2022). Impacts of climate and land-use changes on hydrological processes of the source region of Yellow River, China. *Sustainability, 14*(22), 1.

Jia, Y., Wang, H., Zhou, Z., Qiu, Y., Luo, X., Wang, J., Yan, D., & Qin, D. (2006). Development of the WEP-L distributed hydrological model and dynamic assessment of water resources in the Yellow River basin. *Journal of Hydrology, 331*(3–4), 606–629.

Jiao, S., Zhang, M., Wang, Y., Liu, J., & Li, Y. (2014). Variation of soil nutrients and particle size under different vegetation types in the Yellow River Delta. *Acta Ecologica Sinica, 34*(3), 148–153.

Jin, X., Jin, H., Luo, D., Sheng, Y., Wu, Q., Wu, J., Wang, W., Huang, S., Li, X., Liang, S., & Wang, Q. (2022). Impacts of permafrost degradation on hydrology and vegetation in the source area of the Yellow River on Northeastern Qinghai-Tibet Plateau, Southwest China. *Frontiers in Earth Science, 10*, 845824.

Kusky, T., Windley, B. F., & Zhai, M. G. (2007). Tectonic evolution of the North China Block: from orogen to craton to orogen. *Geological Society, London, Special Publications, 280*(1), 1–34.

Lan, Y. C., Kang, E. S., Ma, Q. J., Yang, W. H., & Yao, Z. Z. (1999). Runoff forecast model for inflow to the Longyangxia Reservoir in the upper Yellow River basin during spring. *Journal of Glaciology and Geocryology, 21*(4), 391. (In Chinese).

Lan, H., Peng, J., Zhu, Y., Li, L., Pan, B., Huang, Q., Li, J., & Zhang, Q. (2021). Research on geological and surfacial processes and major disaster effects in the Yellow River Basin. *Science China Earth Sciences, 65*, 1–23.

Lawler, A. (2009). Beyond the yellow river: how China became China. *Science, 325*, 930–935.

Li, J. J., Fang, X. M., Ma, H. Z., Zhu, J. J., Pan, B. T., & Chen, H. L. (1996). Geomorphological and environmental evolution in the upper reaches of the Yellow River during the late Cenozoic. *Science in China, 39*, 380–390.

Li, Q., Yang, M., Wan, G., & Wang, X. (2016). Spatial and temporal precipitation variability in the source region of the Yellow River. *Environmental Earth Sciences, 75*, 1–14.

Li, G., Sun, S., Han, J., Yan, J., Liu, W., Wei, Y., Lu, N., & Sun, Y. (2019a). Impacts of Chinese Grain for Green program and climate change on vegetation in the Loess Plateau during 1982–2015. *Science of the Total Environment, 660*, 177–187.

Li, M., Zhang, S., Xu, Q., Xiao, J., & Wen, R. (2019b). Spatial patterns of vegetation and climate in the North China Plain during the Last Glacial Maximum and Holocene climatic optimum. *Science China Earth Sciences, 62*, 1279–1287.

Lin, A., Yang, Z., Sun, Z., & Yang, T. (2001). How and when did the Yellow River develop its square bend? *Geology, 29*(10), 951–954.

Liu, T. S. (1965). *The loess deposits in China* (p. 224). Science. (In Chinese).

Liu, Y. (2020). Neogene fluvial sediments in the northern Jinshaan Gorge, China: Implications for early development of the Yellow River since 8 Ma and its response to rapid subsidence of the Weihe-Shanxi Graben. *Palaeogeography, Palaeoclimatology, Palaeoecology, 546*, 109675.

Liu, Y., Miao, H. T., Huang, Z., Cui, Z., He, H., Zheng, J., Han, F., Chang, X., & Wu, G. L. (2018). Soil water depletion patterns of artificial forest species and ages on the Loess Plateau (China). *Forest Ecology and Management, 417*, 137–143.

Liu, B., Pan, L., Qi, Y., Guan, X., & Li, J. (2021). Land use and land cover change in the Yellow River Basin from 1980 to 2015 and its impact on the ecosystem services. *Land, 10*(10), 1080.

Loveland, T. R., & Belward, A. S. (1997). The international geosphere biosphere programme data and information system global land cover data set (DISCover). *Acta Astronautica, 41*(4–10), 681–689.

Na, L., Genxu, W., Guangsheng, L., Yun, L., & Xiangyang, S. (2013). The ecological implications of land use change in the Source Regions of the Yangtze and Yellow Rivers, China. *Regional Environmental Change, 13*, 1099–1108.

Nie, J., Stevens, T., Rittner, M., Stockli, D., Garzanti, E., Limonta, M., Bird, A., Andò, S., Vermeesch, P., Saylor, J., Lu, H., Breecker, D., Hu, X., Liu, S., Resentini, A., Vezzoli, G., Peng, W., Carter, A., Ji, S., & Pan, B. (2015). Loess Plateau storage of Northeastern Tibetan Plateau-derived Yellow River sediment. *Nature Communication, 6*, 68511.

Qiu, J. (2008). China: the third pole. *Nature, 454*(24), 393–396.

Ran, L., & Lu, X. X. (2012). Delineation of reservoirs using remote sensing and their storage estimate: an example of the Yellow River basin, China. *Hydrological Processes, 26*(8), 1215–1229.

Ren, M. E., & Zhu, X. M. (1994). Anthropogenic influences on changes in the sediment load of the Yellow River, China, during the Holocene. *The Holocene, 4*(3), 314–320.

Royden, L. H., Burchfiel, B. C., & van der Hilst, R. D. (2008). The geological evolution of the Tibetan Plateau. *Science, 321*(5892), 1054–1058.

Song, X., Yang, G., Yan, C., Duan, H., Liu, G., & Zhu, Y. (2009). Driving forces behind land use and cover change in the Qinghai-Tibetan Plateau: a case study of the source region of the Yellow River, Qinghai Province, China. *Environmental Earth Sciences, 59*, 793–801.

Wang, G. X., Ju, Q., Guodong, C., & Yuanmin, L. (2002). Soil organic carbon pool of grassland soils on the Qinghai-Tibetan Plateau and its global implication. *Science of the Total Environment, 291*(1–3), 207–217.

Wang, G., Ding, Y., & Wang, J. (2004). Land ecological changes and evolutional patterns in the source regions of the Yangtze and Yellow Rivers in recent 15 years. *Acta Geographica Sinica-Chinese Edition, 59*(2), 163–173.

Wang, H., Guo, Z. G., Xu, X. H., Liang, T. G., & Ren, J. Z. (2007). Response of vegetation and soils to desertification of alpine meadow in the upper basin of the Yellow River, China. *New Zealand Journal of Agricultural Research, 50*(4), 491–501.

Wang, X., Yu, J., Zhou, D., Dong, H., Li, Y., Lin, Q., Guan, B., & Wang, Y. (2012). Vegetative ecological characteristics of restored reed (Phragmites australis) wetlands in the Yellow River Delta, China. *Environmental Management, 49*, 325–333.

Wang, H., Sun, J., Li, W., Wu, J., Chen, Y., & Liu, W. (2016). Effects of soil nutrients and climate factors on belowground biomass in an alpine meadow in the source region of the Yangtze-Yellow rivers, Tibetan Plateau of China. *Journal of Arid Land, 8*, 881–889.

Wang, R., Dong, Z., & Zhou, Z. (2020). Different responses of vegetation to frozen ground degradation in the source region of the Yellow River from 1980 to 2018. *Chinese Geographical Science, 30*, 557–571.

Wei, Y., Jiao, J., Zhao, G., Zhao, H., He, Z., & Mu, X. (2016). Spatial–temporal variation and periodic change in streamflow and suspended sediment discharge along the mainstream of the Yellow River during 1950-2013. *Catena, 140*, 105–115.

Wohlfart, C., Kuenzer, C., Chen, C., & Liu, G. (2016). Social–ecological challenges in the Yellow River basin (China): a review. *Environmental Earth Sciences, 75*, 1–20.

Wu, B., Zhaoyin, W., & Changzhi, L. (2004). Yellow River Basin management and current issues. *Journal of Geographical Sciences, 14*, 29–37.

Wu, R. Z., Xiao, H., Lu, G., & Chen, J. (2015). Assessment of climate change effects on water resources in the Yellow River basin, China. *Advances in Meteorology, 2015*, 1–8.

Wu, X., Wang, S., Fu, B., Feng, X., & Chen, Y. (2019). Socio-ecological changes on the Loess Plateau of China after Grain to Green Program. *Science of the Total Environment, 678*, 565–573.

Xiao, J. (2014). Satellite evidence for significant biophysical consequences of the "Grain for Green" Program on the Loess Plateau in China. *Journal of Geophysical Research: Biogeosciences, 119*(12), 2261–2275.

Xiao, G., Sun, Y., Yang, J., Yin, Q., Dupont-Nivet, G., Licht, A., Kehew, A. E., Hu, Y., Geng, J., Dai, G., & Zhao, Q. (2020). Early Pleistocene integration of the Yellow River I: detrital-zircon evidence from the North China Plain. *Palaeogeography, Palaeoclimatology, Palaeoecology, 546*, 109691.

Xiao, L., Zhao, M., Liu, G., Li, P., Liu, F., & Xue, S. (2022). Soil biochemical index-based assessment of the effect of drought stress on the rhizosphere soil quality in three typical grass species in the Loess Plateau, China. *Journal of Soils and Sediments, 22*(12), 2982–2994.

Xu, Z. X., Takeuchi, K., Ishidaira, H., & Zhang, X. W. (2002). Sustainability analysis for Yellow River water resources using the system dynamics approach. *Water Resources Management, 16*, 239–261.

Xu, M., Wang, G., Wang, Z., Hu, H., Singh, D. K., & Tian, S. (2022). Temporal and spatial hydrological variations of the Yellow River in the past 60 years. *Journal of Hydrology, 609*, 127750.

Xue, C. (1993). Historical changes in the Yellow River delta, China. *Marine Geology, 113*(3–4), 321–330.

Yang, W., Shao, M. A., Peng, X., & Xia, W. (1999). On the relationship between environmental aridization of the Loess Plateau and soil water in loess. *Science in China Series D: Earth Sciences, 42*(3), 240–249.

YRCC. (2013). *Planning report of the Yellow River Basin [in Chinese]*. Yellow River Water Conser.

Yu, L. (2006). The Huanghe (Yellow) River: recent changes and its countermeasures. *Continental Shelf Research, 26*(17–18), 2281–2298.

Yu, Y., Wang, H., Shi, X., Ran, X., Cui, T., Qiao, S., & Liu, Y. (2013). New discharge regime of the Huanghe (Yellow River): Causes and implications. *Continental Shelf Research, 69*, 62–72.

Yuan, X., Sheng, X., Chen, L., Tang, Y., Li, Y., Jia, Y., Qu, D., Wang, Q., Ma, Q., & Zuo, J. (2022). Carbon footprint and embodied carbon transfer at the provincial level of the Yellow River Basin. *Science of the Total Environment, 803*, 149993.

Yue, L., Li, J., Zheng, G., & Li, Z. (2007). Evolution of the Ordos Plateau and environmental effects. *Science in China Series D: Earth Sciences, 50*(2), 19–26.

Zhang, J. T., Ru, W., & Li, B. (2006). Relationships between vegetation and climate on the Loess Plateau in China. *Folia Geobotanica, 41*, 151–163.

Zhang, Y., Huang, M., & Lian, J. (2015). Spatial distributions of optimal plant coverage for the dominant tree and shrub species along a precipitation gradient on the central Loess Plateau. *Agricultural and Forest Meteorology, 206*, 69–84.

Zhang, X., Zhao, W., Liu, Y., Fang, X., & Feng, Q. (2016). The relationships between grasslands and soil moisture on the Loess Plateau of China: A review. *Catena, 145*, 56–67.

Zhang, X. B., Liu, Y., Wang, S. J., Liu, W. M., & Xue, W. X. (2018). On the chronology of the Yellow Rivers and the Yangtze Rivers. *Mountain Research, 36*(5), 661–668.

Zhang, Y., Hassan, M. A., King, L., Fu, X., Istanbulluoglu, E., & Wang, G. (2020). Morphometrics of China's Loess Plateau: The spatial legacy of tectonics, climate, and loess deposition history. *Geomorphology, 354*, 107043.

Zhang, H., Lu, H., Zhou, Y., Cui, Y., He, J., Lv, H., Wang, K., & Wang, X. (2021). Heavy mineral assemblages and UPb detrital zircon geochronology of sediments from the Weihe and Sanmen Basins: New insights into the Pliocene-Pleistocene evolution of the Yellow River. *Palaeogeography, Palaeoclimatology, Palaeoecology, 562*, 110072.

Zhao, J., Luo, X., Ma, Y., Shao, T., & Yue, Y. (2017). Soil characteristics and new formation model of loess on the Chinese Loess Plateau. *Geosciences Journal, 21*, 607–616.

Zheng, H., Zhang, L., Liu, C., Shao, Q., & Fukushima, Y. (2007). Changes in stream flow regime in headwater catchments of the Yellow River basin since the 1950s. *Hydrological Processes: An International Journal, 21*(7), 886–893.

Zhou, H., Han, S., & Liu, W. (2020). Evaluation of two generalized complementary functions for annual evaporation estimation on the Loess Plateau, China. *Journal of Hydrology, 587*, 124980.

Zhu, Z. Y. (1989). The formation of river terraces and evolution of drainage system in the middle Yellow River. *Acta Geographica Sinica, 44*(4), 429–440.

Chapter 3
Paleo-Climates and Environments

Contents

3.1 Introduction

Climate is one of the most important variables in the environmental systems. It impacts the biological community (Graham & Grimm, 1990; Smol et al., 2005; Thomas, 2011; Pi et al., 2021), soil formation (Jenny, 1994; Van Breemen & Buurman, 2002), geomorphic and hydrological processes (Bull, 1991; Gutierrez & Benito, 2005; McLaughlin et al., 2017; Juracek & Fitzpatrick, 2022), as well as the landscape development (Grunewald & Scheithauer, 2010). An understanding of the climate in the past holds a key to the prediction of future climate changes as the Earth's orbital cycles play a critical role in climate evolution (Zachos et al., 2001). The paleoclimates in the Yellow River basin have been complicated by the regional settings, with the uplift of the Tibet plateau and the formation of the Asian monsoon. Consequently, thick loesses deposited in the vast area of the middle section of the

T. Zhu, *Watershed Environmental Degradation and Sustainable Development*,
https://doi.org/10.1007/978-3-032-02448-0_3

basin. The climates, together with geologic and geomorphic conditions, make the Yellow River Basin ecological sensitive, which has been more vulnerable since the human settlement in the region. Determination of the natural erosion in the region prior to the human settlement can help us evaluate the contribution of human activities to the current status of erosion. The information retrieved on palaeoecological response to the paleoclimates is helpful to the restoration of the current degraded environments. In this chapter, a comprehensive review of research will be conducted on the following aspects: (1) the various methods employed in reconstructing the paleoclimates and paleoenvironments in the Yellow River basin, (2) the effect of uplift of the Tibet Plateau on the climates in the region, (3) the deposition of the thick loess in the middle reaches of the Yellow River; (3) the sediment load of the Yellow River and the migration of the lower channels.

3.2 Methods for Reconstruction of Paleoclimates and Paleoenvironments

A variety of methods and techniques have been applied to retrieve information about paleoclimates and paleoenvironments in the Yellow River basin, such as stable isotopes (Li et al., 2020; Hu et al., 2012), loess deposits (Liu, 1985; An et al., 2001; Ding et al., 2002), lake deposits (Pu et al., 2020; Zhao et al., 2021), historical documents (Tan, 1962; Shi, 1999), fluvial landforms (Li et al., 1997; Pan et al., 2009), tree rings (Gou et al., 2007a, 2007b; Wang et al., 2022), among others. Most of these techniques depend on physical, chemical, and biological proxies to reconstruct the history of paleoclimates. However, they often involve assumptions and uncertainties, which can lead to inconsistencies in research findings. The following section examines the principles and applications of each method in the region.

3.2.1 Isotopes Composition in the Deposited Sediments

Among all the isotopes used in paleoclimate reconstructions, oxygen isotopes are the most commonly utilized. Oxygen isotope records have been extensively employed as a proxy to interpret the Asian Monsoon and its effects on the climate of the Yellow River basin (Dettman et al., 2001; Tian et al., 2004; An et al., 2001; Garzione et al., 2000; Wu et al., 2015; Wernicke et al., 2017). This method is based on the variations in the ratio of $^{18}O/^{16}O$ of carbonate sediments. Sediments have been taken from the different lakes in the Tibet region for isotope composition analysis (Lister et al., 1991; Wei & Gasse, 1999; Shen et al., 2005; Zhang et al., 2021; Wu et al., 2022a, 2022b). Usually, the chronology of lake sediments can be determined by short-lived radioactive isotopes (e.g. ^{210}Pb, ^{137}Cs, AMS^{14}C, etc.) (Björck & Wohlfarth, 2002). Deposition rates of lacustrine sediments (Brauer et al., 1999; Halfman & Johnson, 1988), pollen analysis (Bent & Wright Jr, 1963; Martin, 1968; Liu et al., 2002; Ji et al., 2005), water chemistry (Zhang et al., 2004; Dong et al., 2006), total organic content (TOC) (Axford et al., 2011; Kylander et al., 2013), and

$\delta^{18}O$ have been used as sensitive proxies for environment changes over the periods of sediment deposition. For example, high TOC levels indicate a warm environment with high primary productivity whereas low TOC levels suggest a cold environment with low primary productivity. The temporal variability in deposition rates of lake sediments reflect the changes of vegetation cover, erosion processes and sources of sediments that were caused by climate changes.

3.2.2 *Loess Deposits*

The thick, wind-blown loess deposits in the middle reaches of the Yellow River basin provide a continuous record of past climate and environmental changes over the past 2.6 Ma (Heller & Liu, 1982; Liu, 1985; Ding et al., 1998, 1999; Roe, 2009; An et al., 2001). The late Tertiary red clay sediments in the region may be wind-blown in origin and extend the eolian record in the Loess Plateau from 2.6 Ma down to about 7.2 Ma (Ding et al., 1998). Abundant information on regional climatic and environmental changes can be derived from the loess and red earth deposits using a range of proxy indicators. During a dry cold period, dust accumulation rates are high, soil forming time is relatively shorter than a warm wet period when dust accumulation rates are lower that allow ample time for a soil to develop. A paleosol in a loess-paleosol sequence suggests the warm-humid climate with a summer monsoon dominance, while the loess implies the cold-dry climate with a winter monsoon dominance (An et al., 1991). Each paleosol is enriched in nitrogen and organic carbon, depleted of carbonate due to leaching, and displays higher values of pedogenic magnetic susceptibility, frequency-dependent susceptibility (%), remanences and magnetization (Maher et al., 2003). The magnetic susceptibility of the loess-paleosol sequences has been widely used to reconstruct paleoclimate variations corresponding to the loess and paleosol alternations, with high magnetic susceptibility in paleosol units contrasting with low values in loess units (Liu, 1985; Lu et al., 2018; Verosub et al., 1993). However, Ding et al. (1998) found that magnetic susceptibility in the red clay is not as useful as in the loess-soil sequence to indicate summer monsoon variations. Various chemical and physical indexes of loess deposits are also used as climate proxies such as the ratio Rb/Sr and the proportion of coarse particles. Rb is relatively stable element and Sr is mobile element in weathering intensity in soil development. Thus, a higher Rb/Sr may indicate higher degree of weathering associated with stronger summer monsoon conditions whereas a high fraction of coarse particles may indicate a dry and cold climatic condition in the source regions of dust (An et al., 2001).

3.2.3 *Pollen Analysis*

Pollen analysis of lacustrine sediments, bog peat, or ice cores have been conducted in the Tibet Plateau (e.g., Liu et al., 1998; Herzschuh, 2007; Dupont-Nivet et al., 2008), Loess Plateau (Li et al., 2003; Jiang et al., 2013) and the Yellow River delta (Yi et al., 2003). The assemblages of fossil pollen are used to infer the paleo-vegetation and paleoclimates. It is noted that the applicability of pollen ratios should be tested with modern pollen data (Herzschuh, 2007). Such information about modern pollen source areas, production, deposition and preservation, as well as the pollen-vegetation and pollen-climate relationships is critical to the reconstruction of paleoclimate and paleo vegetation (Tang et al., 2021; Xu et al., 2015).

3.2.4 Tree Ring Records

Preserving climate variability in tree-ring records is useful in reconstructing past climate changes (Esper et al., 2002). By calibrating a tree ring width chronology with modern observed climate and hydrological data, tree-ring records have been used to examine the temporal variability of temperature (Gou et al., 2007a), precipitation (Chen et al., 2021),streamflow (Gou et al., 2010;Gou et al., 2007b), and drought conditions (Liang et al., 2006) in the past hundreds or thousands years in the Yellow River basin and Tibet Plateau.

3.2.5 Archaeological Evidence and Historical Documents

The Yellow River basin is the cradle of Chinese civilization. Archaeological evidence indicates that prehistoric human occupation in the basin can date back to nearly 9000 years ago (Zhang et al., 1999). Agricultural societies first emerged approximately 7000 years ago in the region (Zhang et al., 2012). Archaeological investigations and excavations indicate the sequence of those cultures in the Yellow River basin, among which include Majiayao culture (5900–4300 Cal yr. BP and Machang style, 4300–4000 Cal yr. BP), Qijia (4100–3600 Cal yr. BP), Kayue (3600–2600 Cal yr. BP), and Xindian (3400–2700 Cal yr. BP) culture in the upper basin (Dong et al., 2013); Peiligang Culture (9000–7000 Cal yr. BP), Laoguantao Culutre (6000–5000 BC), Yangshao Culture (5000–3000 BC), Longshan Culture (2500–2000 BC), Yingxu (1300–1000 BC), Erlishan Culture (1300–800 BC) in the lower and middle basin. Abundant archaeological relics of prehistoric culture throughout the region have been used to reconstruct the paleoclimate record and paleoenvironments of the region (Zhuang & Kidder, 2014; Zhang et al., 2011; Li et al., 2021; Jia et al., 2022).

The historical records of the Yellow River have been documented since the Spring and Autumn and Warring States Period (770–221 BC). Historical documents have been used to retrieve information on temperature, precipitation, biological community, extreme flood events, channel migrations, as well as human impacts on environments in the basin (Ge et al., 2003; Zheng et al., 2006; Chen et al., 2012; He et al., 2006).

3.3 Rise of the Tibetan Plateau and its Impacts on the Paleoclimates

The timing and amplitude of the Tibet Plateau uplift are in debate and the research progress has been periodically reviewed (Li et al., 1979, 2015; Molnar et al., 2010; Ding et al., 2022). Prior to the collision of India and the Eurasia continents, the Qinghai-Tibet area is believed to be a planation surface (Li et al., 2015). The Tibetan Plateau was formed through the collision of the Indian subcontinent with the Eurasian Plate, leading to the compression and thickening of the Earth's crust. This monumental geological event gave rise to significant mountain ranges such as the Himalayas, Karakoram, and Tien Shan (Royden et al., 2008). Geological evidence suggests that the collision between India and Asian-affinity rocks commenced between 65 and 60 million years ago, based on the dating of the sedimentary rocks in the Indian continental margin (Molnar et al., 2010; Ding et al., 2022). Wang et al. (2011) suggested there are four intensive uplift stages, i.e., 60–35, 25–17 and 12–8 Ma, and since about 5 Ma, based on synthetical analysis of low-temperature thermo-chronology data, sedimentary deposit records, and structural deformation records of different areas. The rapid uplifts in the Tibetan Plateau show great spatial variations. Ding et al. (2022) indicated that the high mountain ranges in southern and central Tibet were created by ~55 Ma and ~ 45 Ma, respectively; The basins separated by the mountain ranges were uplifted further between ~38 and 29 Ma; The northern Tibet plateau and the Himalaya Mountains rapidly rose from <3 km to near-modern elevations at ~20 Ma. However, some people believed the Tibet Plateau underwent a rapid uplift much later. For example, Li et al. (1979, 1996, 2014, 2015) suggest that the rapid persistent rise of the Tibet plateau began ~8 ± 1 Ma followed by stepwise accelerated rise at ~3.6 Ma, 2.6 Ma, 1.8–1.7 Ma, 1.2–0.6 Ma and 0.15 Ma. The Yellow River basin developed at ~1.7 Ma and evolved to its present pattern through stepwise backward-expansion toward its source area in response to the stepwise uplift of the plateau. Tectonically, the Tibetan Plateau is divided into four major sutures from north to south: the Kunlun suture, the Litian Lake-Jinsha River suture, the Bangong Lake-Nu Jiang River suture, and the Indus-Yarlung Zangbo River suture. These sutures delineate various blocks, including the Kunlun, Songpan-Garze, Qangtang, Qamdo, Lhasa, and Himalayan blocks, each playing a distinct role in the plateau's complex tectonic structure (Chang et al., 1989).

The uplift of the Tibet Plateau has profound impacts on the climate and environment at both regional (e.g. the Yellow River basin) and global scale (Li et al., 1979; Held, 1983; Harrison et al., 1992; Shi et al., 1999; An et al., 2001; Molnar et al., 2010). The rising Tibet plateau impacted climates through three mechanisms: (1) block the cool, dry air southward movement and warm, wet air northward movement; (2) change the vertical temperature structure of the atmosphere above the plateau; (3) modify the flow path of the westerly subtropical jet stream. However, the relative importance of each of those mechanisms in affecting climates is still in debate.

The rise of the Tibet Plateau caused the northward expansion of the Asian monsoon climates. Wu et al. (2022a, 2022b) think the Asian Monsoon was limited to the tropical Monsoon prior to the rise of the Tibet Plateau south of 20–22° N, due to the seasonal migration of Intertropical Convergence Zone (ITC). With the progressive rise of the Tibet Plateau, the Asian Monsoon system first expanded northward to the southern subtropical region in East Asia at 41 Ma, and then further expanded to the modern-like monsoon pattern at ~26 Ma, with a boundary between north-west dry to south-east wet along 30–36 °N. Nevertheless, Tang and Liu (1995) think the development of Monsoon was much later, probably happened in the transition of Pliocene to the Quaternary. Zheng et al. (2002) suggests that four glaciations developed in the Quaternary: The Xixiabangma, Naynayxungla, Guxiang, and Baiyu Glaciation, at about 800–1170, 500–720, 130–300, and 10–70 ka, respectively. During the maximum glaciation, the ice coverage could reach at least 500,000 km^2, in comparison to 46,640 km^2 at present (Shi et al., 2000). Chen et al. (2020) reported that during the Holocene period, there was a peak in summer temperatures from the early to mid-Holocene (9–5 ka cal BP), followed by a general cooling trend leading up to the present, based on a compilation of pollen data collected from the region. Correspondingly, the forest reached maximum extent in the mid-Holocene, while alpine steppe has shrunk, and alpine meadow and desert have gradually expanded. The rise of the Tibet Plateau led to the heating of the atmosphere over the region during the spring and summer (Molnar et al., 2010). It further increases the aridity of central Asia (An et al., 2001). It also enhances the interplay between the Asian Monsoon and the Westerlies (Prell et al., 1992).

3.4 Paleo-Climate Change and Paleo-Environment on the Loess Plateau

The central section of the Yellow River basin is covered by the world's thickest loess layers, forming the Loess Plateau. From a geological standpoint, the Loess Plateau is located at the southern edge of the North China (Sino-Korean) Craton. This ancient craton merged with the South China Craton in the early Triassic Period, playing a pivotal role in forming the present-day East Asian landmass (Zhang et al., 2020). Spanning more than 440,000 square kilometers, the loess deposits have an average depth of about 100 meters, though their thickness varies widely from 0 to 350 meters. The origin of these loess deposits, largely attributed to wind-blown sediment from central Asia, dates back to the Quaternary Period (Ding et al., 2002). A complete profile of the loess layers includes the Malan Loess from the late Pleistocene, the Lishi Loess from the mid-Pleistocene, and the Wucheng Loess from the early Pleistocene, as categorized by Liu (1965; 1985). Underneath the loess deposits is the Red Earth (also referred as the Red Clay), which some people believe is the wind-dust deposited in Tertiary as well (Prell et al., 1992; An et al., 2001). But others argued that the Red Earth may not origin from eolian process as its particle sizes are relatively homogeneous and do not show similar spatial variability of wind-blown loess (Li & Fang, 1999). Guo et al. (2001) suggested that the formation of the Red Earth consists of three parts with different formation mechanisms. The Lower part (>6.2 Ma BP) is a water-reworked deposit related to alluvial and slope processes; the Middle part (from ~6.2 to ~3.6 Ma BP) was derived from in situ eolian dust deposition instead of from the central Asian, but was significantly affected by groundwater oscillations; and the Upper part (from ~3.6 to ~2.6 Ma BP) is an eolian formation, similar to the Quaternary loess-soil sequence.

The continuous deposition of loess and paleosols provides an excellent record of paleoclimates and environmental conditions in the Quaternary. Numerous studies have been conducted to retrieve paleoclimate information from the deposits using isotope (Liu, 1985; Liu et al., 1996; Xie, 1999; Gu et al., 2003; Tian et al., 2003; Kaakinen et al., 2006; Xu et al., 2021), pollen (Sun et al., 1997a, 1997b, 2006; Ju & Chen, 1998; Xia et al., 1998; Jiang & Ding, 2005; Hui et al., 2021), particle size (Lu & An, 1998; Chen et al., 2000; Nugteren et al., 2004),paleomagnetic methods (Evans & Heller, 1994; Ding et al., 1999; Maher, 2016), and biomarkers (Bai et al., 2009). The composition of grain sizes in the loess-paleosols sequences is directly related to the paleoclimate changes (Porter & An, 1995). Lu and An (1998) found that the sub-coarse grain fraction is a good proxy index of East Asia winter monsoon strength. During the glacial period, continental ice-sheets expansion and greater sea-ice cover in the Northern Hemisphere greatly strengthened the Siberian High (Ding et al., 1995). Thus, cold and dry winter monsoon dominates and the deposition rates of dusts in the middle reaches of Yellow River from central Asia tended to be high, with greater fraction of coarse grains. In the interglacial periods, summer wet and warm Monsoon prevails, which results in higher fraction of fine grains and lower deposition rates, often associated with paleosols. Spatially, the particle size of loess decreases from northwest to southeast in the Loess Plateau region. In other words, the farther away from the source region of the dust, the finer the deposited loess particles. The climate was relatively stable and less variable during the 8.1–2.6 Ma period (Bai et al., 2009). However, the drying and cooling climates occurred at about 2.4 Ma, 1.2 Ma, and 0.5 Ma, while less pronounced shifts took place at 1.65 Ma, 0.8 Ma, and 0.2 Ma (Kukla & An, 1989; Ding et al., 2002; Liang et al., 2018).

A number of studies have explored the spatial variation of paleovegetation on the Loess Plateau, as comprehensively reviewed by Lü et al. (2003) and Xu et al. (2021). Pollen analyses from loess deposits suggest that herbs were predominant during the early to mid-Holocene, with species such as *Artemisia*, *Echinops*-type, *Taraxacum*-type, and members of the Chenopodiaceae family being the most common (Jiang et al., 2013). According to Lü et al. (2003), the paleo vegetation in the Loess Plateau exhibited significant spatial variability. Forests primarily covered the mountainous and valley regions, with coniferous forests dominating the northern areas and a mix of broadleaf and coniferous forests in the south. In contrast, prairies and steppes were more common in the flatter areas. The vegetation also responded to climatic shifts over time. As climates transitioned from warm and humid to cold and dry periods, vegetation types would shift accordingly, such as prairie transforming into steppe, evergreen forests into coniferous forests, and steppes into semi-deserts, and vice versa. However, there is still an ongoing debate regarding the natural vegetation in the region (Sun et al., 1997a, 1997b; Shi, 1999; Kaakinen et al., 2006; Zhang et al., 2008; Jiang et al., 2013; Liu et al., 2018; Wang et al., 2018). Vegetation changes in recent history due to human activities have been identified through the analysis of historical documents (Tan, 1962; Shi, 1991), which will be explored in Chap. 7.

3.5 Paleoclimates of the Lower Yellow River Basin and Channel Evolution

The lower Yellow River flows eastward on the North China Plain and discharges to the Bohai sea now. The timing of the lower Yellow River formation is still in debate. The Yellow River cut through the upland and formed the Sanmenxia Gorge, flowing towards the sea. Hu et al. (2017) suggested that the integration of the Yellow River with the upper and middle channel systems occurred between 3.6–1.2 Ma. Yang et al. (2022), based on the borehole record, deduced that the Yellow River cut through the Sanmen Gorge at 1.6 Ma. But some others suggested the integration happened much later, 0.25 to 0.15 Ma (Wang et al., 2002; Jiang et al., 2007; Hu et al., 2019). Prior to the integration, there were a series of paleo-lakes in the middle reaches of the Yellow River, including the largest paleo Sanmen Lake in the region.

The climates in the lower Yellow River basin have generally become colder and drier since 5.6 Ma. Being adjacent to the ocean, the region experiences the regression and transgression in the Quaternary period due to sea level rise and fall. The pollen assemblages and tree rings record show that broadleaf and coniferous mixed forests dominated in the mountains and hills, intrazonal grassland dominated on the alluvial fans, and lake swamp-flood plain developed in the hinterland of the North China Plain in the mid-Holocene, with a mean annual temperature of 3.5 ℃ warmer than present (Zhang et al., 2022;). After human settlements, cultivation of flood plains has been dominated, although it has been frequently disrupted by floods. Qin et al. (2019) believe that the environment deterioration (e.g. flood, dry and cold conditions) was a driving force of initial agricultural intensification as it increased risks of crop failure and famine, and thus induced intensifying agriculture to feed people. This environmental pressure was aggravated by social/political factors and resulted in the further development of agricultural intensification.

The lower Yellow River is extremely prone to flooding. Historical documents recorded more than 1000 floods over the past 4000 years. The main channel has avulsed 26 times in the past 2200 years and has followed its current course since 1855 (Chen et al., 2012; Kong et al., 2020). According to Ren's study (2015), sediment load of the Yellow River increased from 0.5 x 10^8 t/a over the period of 150–8 kBP, with very little human impact to 16 × 10^8 t/a in the period from 1800 to 1970 when the rapid increases of human population and large-scale land farming occurred.

This means that human activities has increased natural erosion by more than 30-fold and the effect of climate changes on the sediment load in the Yellow River was relatively insignificant in comparison to the human impacts. Especially since late Tang Dynasty (618–907 AD), large scale cultivation of steep lands has replaced the nomadic life styles in the middle reaches of the Yellow River, leading to great destruction of natural vegetation. As a result, the sediment load of the Yellow River increased from 2×10^{8}t/a to $16x10^{8}$t/a. Before then, the sediment load fluctuated between 0.5 to 2×10^{8}t/a. According to Ren's studies, about 72% of the total sediment deposited in the North China Plain, and the remaining escaped to the sea (Ren, 2015).

3.6 Limitations of the Previous Research and Recommendations for Future Study

The uplift of the Tibetan Plateau and its effects on the paleoclimates and paleoenvironments of the plateau and surrounding regions has become a major area of research, attracting scientists from a variety of disciplines around the world. Despite numerous publications, there is still no agreement on the timing, stages of uplift, or the spatial variation of elevation changes across the plateau (Murphy et al., 1997; Molnar et al., 2010; Sun et al., 2014). The timing of the integration of the upper, middle and lower channels of the Yellow River remains a point of disagreement among researchers as well (Lin et al., 2001; Yang et al., 2001; Jiang et al., 2007; Stevens et al., 2013). It is widely accepted that the uplift of the Tibetan Plateau has had profound effects on the climate, influencing systems like the East Asian Monsoon and contributing to the aridity in the upper and middle reaches of the Yellow River basin. However, the paleoclimate history of both the Tibetan Plateau and the Yellow River basin remains a topic of debate (Wang et al., 2008; Song et al., 2022). Additionally, the mechanisms driving paleoclimate changes due to the uplift of the Tibetan Plateau are not yet fully understood (Yuan et al., 2004; Molnar et al., 2010).

The uncertainties and conflicting results mentioned above are partly due to the research methodologies and the assumptions underlying them. When using sediment deposits to reconstruct paleoclimates, the number and distribution of lakes are crucial. Relying on data from only one or a few locations and then extrapolating to larger regions can lead to significant errors in reconstructing paleoclimate history. Furthermore, sediment deposits in lakes are influenced by various factors, including lake size, whether the lake is open or closed, depth, the surrounding catchment environment, and atmospheric circulation patterns. Understanding the hydrological and biochemical processes of individual lakes is crucial when using isotope proxies in sediments. The isotope method requires precise calibration of modern lake environments to establish a relationship between current climate conditions and isotope compositions. Another issue with lacustrine deposits is the uncertainty of sediment ages determined by radioactive techniques (Hou et al., 2012). More studies should focus on combining multiple proxies recorded in lacustrine sediments within a single lake, along with paleoclimate records from lakes across the region, to better interpret the paleoenvironments of the area. Reconstructed paleoclimate records obtained through different techniques should be cross-validated and compared with other regional and global paleoclimate records.

The paleoclimate information obtained from the thick loess profiles is limited to only a few locations in the Loess Plateau (An et al., 1990; Ding et al., 1993). Considering the vast volume of loess deposits in the region, the influence of the source area is likely substantial. Currently, it is assumed that the loess source region is uniform, and variations in loess profiles are attributed solely to paleoclimate conditions. However, the actual spatial extent of the source region is unclear, as is the extent to which spatial and vertical variation in source materials affects the physical and chemical properties of the deposits. Furthermore, loess particle size may change over time due to soil development processes. Compared to studies on paleoclimate information derived from loess deposits, significantly fewer studies have focused on the red earth deposits underlying the loess. Further research is needed to understand the mechanisms behind red earth formation and to utilize red earth deposits for reconstructing paleoclimate history prior to the loess deposits.

Finally, while nearly all previous studies emphasize the importance of understanding past climates for predicting future climate and environmental changes, very few have incorporated paleoclimate research into climate change predictions. Additionally, there is typically minimal integration—if any—of paleoenvironmental studies into the strategies for restoring the currently degraded environments in the region.

References

An, Z. S., Liu, D. S. L., Yanchou, L., Porter, S. C., Kukla, G. H. W. X., Xihao, W., & Yingming, H. (1990). The long-term paleomonsoon variation recorded by the loess-paleosol sequence in Central China. *Quaternary International, 7*, 91–95.

An, Z., Kukla, G., Porter, S. C., & Xiao, J. (1991). Late quaternary dust flow on the Chinese loess plateau. *Catena, 18*(2), 125–132.

An, Z., Kutzbach, J., Prell, W., & Stephen, C. (2001). Evolution of Asian monsoons and phased uplift of the Himalaya–Tibetan plateau since Late Miocene times. *Nature, 411*, 62–66.

Axford, Y., Andresen, C. S., Andrews, J. T., Belt, S. T., Geirsdóttir, Á., Massé, G., Miller, G. H., Ólafsdóttir, S., & Vare, L. L. (2011). Do paleoclimate proxies agree? A test comparing 19 late Holocene climate and sea-ice reconstructions from Icelandic marine and lake sediments. *Journal of Quaternary Science, 26*(6), 645–656.

Bai, Y., Fang, X., Nie, J., Wang, Y., & Wu, F. (2009). A preliminary reconstruction of the paleoecological and paleoclimatic history of the Chinese loess plateau from the application of biomarkers. *Palaeogeography, Palaeoclimatology, Palaeoecology, 271*(1–2), 161–169.

Bent, A. M., & Wright, H. E., Jr. (1963). Pollen analyses of surface materials and lake sediments from the Chuska Mountains, New Mexico. *Geological Society of America Bulletin, 74*(4), 491–500.

Björck, S., & Wohlfarth, B. (2002). 14 C chronostratigraphic techniques in paleolimnology. In *Tracking environmental change using lake sediments* (pp. 205–245). Springer.
Brauer, A., Endres, C., Günter, C., Litt, T., Stebich, M., & Negendank, J. F. (1999). High resolution sediment and vegetation responses to younger Dryas climate change in varved lake sediments from Meerfelder Maar, Germany. *Quaternary Science Reviews, 18*(3), 321–329.
Bull, W. B. (1991). *Geomorphic responses to climatic change*. Department of Energy.
Chang, C. F., Pan, Y. S., & Sun, Y. Y. (1989). The tectonic evolution of Qinghai-Tibet plateau: A review. In *Tectonic evolution of the Tethyan region* (pp. 415–476). Springer.
Chen, F., Feng, Z., & Zhang, J. (2000). Loess particle size data indicative of stable winter monsoons during the last interglacial in the western part of the Chinese loess plateau. *Catena, 39*(4), 233–244.
Chen, Y., Syvitski, J. P., Gao, S., Overeem, I., & Kettner, A. J. (2012). Socio-economic impacts on flooding: A 4000-year history of the Yellow River, China. *Ambio, 41*(7), 682–698.
Chen, F., Zhang, J., Liu, J., Cao, X., Hou, J., Zhu, L., Xu, X., Liu, X., Wang, M., Wu, D., & Huang, L. (2020). Climate change, vegetation history, and landscape responses on the Tibetan plateau during the Holocene: A comprehensive review. *Quaternary Science Reviews, 243*, 106444.
Chen, Y., Chen, F., & Zhang, H. (2021). A tree-ring-based precipitation reconstruction since 1760 CE from northeastern Tibetan plateau, China. *Atmosphere, 12*(4), 416.
Dettman, D. L., Kohn, M. J., Quade, J., Ryerson, F. J., Ojha, T. P., & Hamidullah, S. (2001). Seasonal stable isotope evidence for a strong Asian monsoon throughout the past 10.7 my. *Geology, 29*(1), 31–34.
Ding, Z. L., Rutter, N., & Tungsheng, L. (1993). Pedostratigraphy of Chinese loess deposits and climatic cycles in the last 2.5 Myr. *Catena, 20*(1–2), 73–91.
Ding, Z., Liu, T., Rutter, N. W., Yu, Z., Guo, Z., & Zhu, R. (1995). Ice-volume forcing of east Asian winter monsoon variations in the past 800,000 years. *Quaternary Research, 44*(2), 149–159.
Ding, Z. L., Sun, J. M., Yang, S. L., & Liu, T. S. (1998). Preliminary magnetostratigraphy of a thick eolian red clay-loess sequence at Lingtai, the Chinese loess plateau. *Geophysical Research Letters, 25*(8), 1225–1228.
Ding, Z. L., Xiong, S. F., Sun, J. M., Yang, S. L., Gu, Z. Y., & Liu, T. S. (1999). Pedostratigraphy and paleomagnetism of a~ 7.0 Ma eolian loess–red clay sequence at Lingtai, loess plateau, north-Central China and the implications for paleomonsoon evolution. *Palaeogeography, Palaeoclimatology, Palaeoecology, 152*(1–2), 49–66.
Ding, Z. L., Derbyshire, E., Yang, S. L., Yu, Z. W., Xiong, S. F., & Liu, T. S. (2002). Stacked 2.6-Ma grain size record from the Chinese loess based on five sections and correlation with the deep-sea δ18O record. *Paleoceanography, 17*(3), 5–1.
Ding, L., Kapp, P., Cai, F., Garzione, C. N., Xiong, Z., Wang, H., & Wang, C. (2022). Timing and mechanisms of Tibetan plateau uplift. *Nature Reviews Earth & Environment, 3*, 1–16.
Dong, H., Zhang, G., Jiang, H., Yu, B., Chapman, L. R., Lucas, C. R., & Fields, M. W. (2006). Microbial diversity in sediments of saline Qinghai Lake, China: Linking geochemical controls to microbial ecology. *Microbial Ecology, 51*(1), 65–82.
Dong, G., Jia, X., Elston, R., Chen, F., Li, S., Wang, L., Cai, L., & An, C. (2013). Spatial and temporal variety of prehistoric human settlement and its influencing factors in the upper Yellow River valley, Qinghai Province, China. *Journal of Archaeological Science, 40*(5), 2538–2546.
Dupont-Nivet, G., Hoorn, C., & Konert, M. (2008). Tibetan uplift prior to the Eocene-Oligocene climate transition: Evidence from pollen analysis of the Xining Basin. *Geology, 36*(12), 987–990.
Esper, J., Cook, E. R., & Schweingruber, F. H. (2002). Low-frequency signals in long tree-ring chronologies for reconstructing past temperature variability. *Science, 295*(5563), 2250–2253.
Evans, M. E., & Heller, F. (1994). Magnetic enhancement and palaeoclimate: Study of a loess/palaeosol couplet across the loess plateau of China. *Geophysical Journal International, 117*(1), 257–264.

Garzione, C. N., Quade, J., DeCelles, P. G., & English, N. B. (2000). Predicting paleoelevation of Tibet and the Himalaya from $\delta^{18}O$ vs. altitude gradients of meteoric water across the Nepal Himalaya. *Earth and Planetary Science Letters, 183*, 215–229.

Ge, Q., Zheng, J., Fang, X., Man, Z., Zhang, X., Zhang, P., & Wang, W. C. (2003). Winter half-year temperature reconstruction for the middle and lower reaches of the Yellow River and Yangtze River, China, during the past 2000 years. *The Holocene, 13*(6), 933–940.

Gou, X., Chen, F., Jacoby, G., Cook, E., Yang, M., Peng, J., & Zhang, Y. (2007a). Rapid tree growth with respect to the last 400 years in response to climate warming, northeastern Tibetan plateau. *International Journal of Climatology: A Journal of the Royal Meteorological Society, 27*(11), 1497–1503.

Gou, X., Chen, F., Cook, E., Jacoby, G., Yang, M., & Li, J. (2007b). Streamflow variations of the Yellow River over the past 593 years in western China reconstructed from tree rings. *Water Resources Research, 43*(6), 1.

Gou, X., Deng, Y., Chen, F., Yang, M., Fang, K., Gao, L., Yang, T., & Zhang, F. (2010). Tree ring based streamflow reconstruction for the upper Yellow River over the past 1234 years. *Chinese Science Bulletin, 55*(36), 4179–4186.

Graham, R. W., & Grimm, E. C. (1990). Effects of global climate change on the patterns of terrestrial biological communities. *Trends in Ecology & Evolution, 5*(9), 289–292.

Grunewald, K., & Scheithauer, J. (2010). *Landscape development and climate change in Southwest Bulgaria (Pirin Mountains)*. Springer Science & Business Media.

Gu, Z., Liu, Q., Xu, B., Han, J., Yang, S., Ding, Z., & Liu, T. (2003). Climate as the dominant control on C 3 and C 4 plant abundance in the loess plateau: Organic carbon isotope evidence from the last glacial-interglacial loess-soil sequences. *Chinese Science Bulletin, 48*, 1271–1276.

Guo, Z. T., Peng, S. Z., Hao, Q. Z., Biscaye, P. E., & Liu, T. S. (2001). Origin of the Miocene–Pliocene red-earth formation at Xifeng in Northern China and implications for paleoenvironments. *Palaeogeography, Palaeoclimatology, Palaeoecology, 170*(1–2), 11–26.

Gutierrez, M., & Benito, G. (2005). *Climatic geomorphology*. Elsevier.

Halfman, J. D., & Johnson, T. C. (1988). High-resolution record of cyclic climatic change during the past 4 ka from Lake Turkana, Kenya. *Geology, 16*(6), 496–500.

Harrison, T. M., Copeland, P., Kidd, W. S. F., & Yin, A. N. (1992). Raising Tibet. *Science, 255*(5052), 1663–1670.

He, X., Zhou, J., Zhang, X., & Tang, K. (2006). Soil erosion response to climatic change and human activity during the quaternary on the loess plateau, China. *Regional Environmental Change, 6*(1), 62–70.

Held, I. M. (1983). Stationary and quasi-stationary eddies in the extratropical troposphere: Theory. In B. Hoskins & R. Pearce (Eds.), *Large-scale dynamical processes in the atmosphere* (pp. 127–168). Academic.

Heller, F., & Liu, D. S. (1982). Magnetostratigraphical dating of loess deposits in China. *Nature, 300*(5891), 431–433.

Herzschuh, U. (2007). Reliability of pollen ratios for environmental reconstructions on the Tibetan plateau. *Journal of Biogeography, 34*(7), 1265–1273.

Hou, J., D'Andrea, W. J., & Liu, Z. (2012). The influence of 14C reservoir age on interpretation of paleolimnological records from the Tibetan plateau. *Quaternary Science Reviews, 48*, 67–79.

Hu, B., Li, G., Li, J., Bi, J., Zhao, J., & Bu, R. (2012). Provenance and climate change inferred from Sr–Nd–Pb isotopes of late quaternary sediments in the Huanghe (Yellow River) Delta, China. *Quaternary Research, 78*(3), 561–571.

Hu, Z., Pan, B., Bridgland, D., Vandenberghe, J., Guo, L., Fan, Y., & Westaway, R. (2017). The linking of the upper-middle and lower reaches of the Yellow River as a result of fluvial entrenchment. *Quaternary Science Reviews, 166*, 324–338.

Hu, Z., Li, M., Dong, Z., Guo, L., Bridgland, D., Pan, B., Li, X., & Liu, X. (2019). Fluvial entrenchment and integration of the Sanmen gorge, the lower Yellow River. *Global and Planetary Change, 178*, 129–138.

Hui, Z., Ran, M., Li, H., Liu, C., Guo, B., Zhang, J., Peng, T., Liu, D., & Pan, Y. (2021). Early Pleistocene pollen record from the western Chinese loess plateau and its implications for the evolution of the east Asian summer monsoon. *Science of the Total Environment, 761*, 143304.
Jenny, H. (1994). *Factors of soil formation: A system of quantitative pedology*. Courier Corporation.
Ji, S., Xingqi, L., Sumin, W., & Matsumoto, R. (2005). Palaeoclimatic changes in the Qinghai Lake area during the last 18,000 years. *Quaternary International, 136*(1), 131–140.
Jia, Y. N., Zhang, Y., Huang, C. C., Wang, N., Qiu, H., Wang, H., Xiao, Q., Chen, D., Lin, X., Zhu, Y., & Fu, L. (2022). Late Pleistocene-Holocene aeolian loess-paleosol sections in the Yellow River source area on the northeast Tibetan plateau: Chronostratigraphy, sediment provenance, and implications for paleoclimate reconstruction. *Catena, 208*, 105777.
Jiang, H., & Ding, Z. (2005). Temporal and spatial changes of vegetation cover on the Chinese loess plateau through the last glacial cycle: Evidence from spore-pollen records. *Review of Palaeobotany and Palynology, 133*(1–2), 23–37.
Jiang, F., Fu, J., Wang, S., Sun, D., & Zhao, Z. (2007). Formation of the Yellow River, inferred from loess–palaeosol sequence in Mangshan and lacustrine sediments in Sanmen Gorge, China. *Quaternary International, 175*(1), 62–70.
Jiang, W., Cheng, Y., Yang, X., & Yang, S. (2013). Chinese loess plateau vegetation since the last glacial maximum and its implications for vegetation restoration. *Journal of Applied Ecology, 50*(2), 440–448.
Ju, T. Z., & Chen, X. L. (1998). Research of vegetational succession on the Xiaolongshan Mt. since middle. *Holocene, Acta Botanica Boreali-Occidentalia Sinica (in Chinese), 18*(2), 292–299.
Juracek, K. E., & Fitzpatrick, F. A. (2022). Geomorphic responses of fluvial systems to climate change: A habitat perspective. *River Research and Applications, 38*, 757–775.
Kaakinen, A., Sonninen, E., & Lunkka, J. P. (2006). Stable isotope record in paleosol carbonates from the Chinese loess plateau: Implications for late Neogene paleoclimate and paleovegetation. *Palaeogeography, Palaeoclimatology, Palaeoecology, 237*(2–4), 359–369.
Kong, D., Latrubesse, E. M., Miao, C., & Zhou, R. (2020). Morphological response of the lower Yellow River to the operation of Xiaolangdi dam, China. *Geomorphology, 350*, 106931.
Kukla, G., & An, Z. (1989). Loess stratigraphy in Central China. *Palaeogeography, Palaeoclimatology, Palaeoecology, 72*, 203–225.
Kylander, M. E., Klaminder, J., Wohlfarth, B., & Löwemark, L. (2013). Geochemical responses to paleoclimatic changes in southern Sweden since the late glacial: The Hässeldala port lake sediment record. *Journal of Paleolimnology, 50*(1), 57–70.
Li, J., & Fang, X. (1999). Uplift of the Tibetan plateau and environmental changes. *Chinese Science Bulletin, 44*, 2117–2124.
Li, J. J., Wen, S. X., Zhang, Q. S., et al. (1979). A discussion on the period, amplitude and type of the uplift of the Qinghai-Xizang plateau. *Science in China, Series D, 22*(12), 1314–1328.
Li, J. J., Fang, X. M., Ma, H., et al. (1996). Geomorphological and environmental evolution in the upper reaches of Huanghe River during the late Cenomic. *Science in China Series D-Earth Sciences, 39*, 380.
Li, J. J., Fang, X. M., Van der Voo, R., Zhu, J. J., Niocaill, C. M., Ono, Y., Pan, B. T., Zhong, W., Wang, J. L., Sasaki, T., & Zhang, Y. T. (1997). Magnetostratigraphic dating of river terraces: Rapid and intermittent incision by the Yellow River of the northeastern margin of the Tibetan plateau during the quaternary. *Journal of Geophysical Research: Solid Earth, 102*(B5), 10121–10132.
Li, X., Zhou, J., & Dodson, J. (2003). The vegetation characteristics of the 'Yuan' area at Yaoxian on the loess plateau in China over the last 12 000 years. *Review of Palaeobotany and Palynology, 124*(1–2), 1–7.
Li, J., Fang, X., Song, C., Pan, B., Ma, Y., & Yan, M. (2014). Late Miocene–Quaternary rapid stepwise uplift of the NE Tibetan Plateau and its effects on climatic and environmental changes. *Quaternary Research, 81*(3), 400–423.
Li, J., Zhou, S., Zhao, Z., & Zhang, J. (2015). The Qingzang movement: The major uplift of the Qinghai-Tibetan plateau. *Science China Earth Sciences, 58*(11), 2113–2122.

Li, B., Feng, Q., Wang, F., Wang, X., Li, Z., Zhang, C., Guo, X., Liu, W., & Li, R. (2020). A 1.68 Ma organic isotope record from the Hetao Basin, upper reaches of the Yellow River in northern China: Implications for hydrological and ecological variations. *Global and Planetary Change, 184*, 103061.

Li, K., Gao, W., Wu, L., Hu, H., Gong, P., Li, S., Jin, R., & Si, Y. (2021). Spatial expansion of human settlement during the Longshan period (~ 4.5–~ 3.9 ka BP) and its Hydroclimatic contexts in the lower Yellow River floodplain, eastern China. *Land, 10*(7), 712.

Liang, E., Liu, X., Yuan, Y., Qin, N., Fang, X., Huang, L., Zhu, H., Wang, L., & Shao, X. (2006). The 1920s drought recorded by tree rings and historical documents in the semi-arid and arid areas of northern China. *Climatic Change, 79*(3), 403–432.

Liang, H. Y., Feng, Z. P., Pei, B., Li, Y., & Yang, X. T. (2018). Demographic expansion of two Tamarix species along the Yellow River caused by geological events and climate change in the Pleistocene. *Scientific Reports, 8*(1), 1–11.

Lin, A., Yang, Z., Sun, Z., & Yang, T. (2001). How and when did the Yellow River develop its square bend? *Geology, 29*(10), 951–954.

Lister, G. S., Kelts, K., Zao, C. K., Yu, J. K., & Niessen, K. (1991). Lake Qinghai, China: Closed basin lake levels and the oxygen isotope record for ostracoda since the late Pleistocene. *Palaeogeography Palaeoclimatology, 84*, 141–162.

Liu, T. S. (1985). *Loess and the environment*. China Ocean Press.

Liu, K. B., Yao, Z., & Thompson, L. G. (1998). A pollen record of Holocene climatic changes from the Dunde ice cap, Qinghai-Tibetan Plateau. *Geology, 26*(2), 135–138.

Liu, X., Shen, J., Wang, S., Yang, X., Tong, G., & Zhang, E. (2002). A 16000-year pollen record of Qinghai Lake and its paleo-climate and paleoenvironment. *Chinese Science Bulletin, 47*(22), 1931–1936.

Liu, J., Algeo, T. J., Yang, H., Pan, J., Tang, C., Huang, J., & Xie, S. (2018). Changes in vegetation type on the Chinese loess plateau since 75 ka related to east Asian summer monsoon variation. *Palaeogeography, Palaeoclimatology, Palaeoecology, 510*, 124–139.

Lu, H., & An, Z. (1998). Paleoclimatic significance of grain size of loess-palaeosol deposit in Chinese loess plateau. *Science in China Series D: Earth Sciences, 41*, 626–631.

Lü, H., Liu, D., & Guo, Z. (2003). Natural vegetation of geological and historical periods in loess plateau. *Chinese Science Bulletin, 48*(5), 411–416.

Lu, H., Jia, J., Wang, Y., Yin, Q., & Xia, D. (2018). The cause of extremely high magnetic susceptibility of the S5S1 paleosol in the central Chinese loess plateau. *Quaternary International, 493*, 252–257.

Maher, B. A. (2016). Palaeoclimatic records of the loess/palaeosol sequences of the Chinese loess plateau. *Quaternary Science Reviews, 154*, 23–84.

Maher, B. A., MengYu, H., Roberts, H. M., & Wintle, A. G. (2003). Holocene loess accumulation and soil development at the western edge of the Chinese loess plateau: Implications for magnetic proxies of palaeorainfall. *Quaternary Science Reviews, 22*(5–7), 445–451.

Martin, A. R. H. (1968). Pollen analysis of Groenvlei lake sediments, Knysna (South Africa). *Review of Palaeobotany and Palynology, 7*(2), 107–144.

McLaughlin, B. C., Ackerly, D. D., Klos, P. Z., Natali, J., Dawson, T. E., & Thompson, S. E. (2017). Hydrologic refugia, plants, and climate change. *Global Change Biology, 23*(8), 2941–2961.

Molnar, P., Boos, W. R., & Battisti, D. S. (2010). Orographic controls on climate and paleoclimate of Asia: Thermal and mechanical roles for the Tibetan plateau. *Annual Review of Earth and Planetary Sciences, 38*(1), 77–102.

Murphy, M. A., Yin, A., Harrison, T. M., Durr, S. B., Ryerson, F. J., & Kidd, W. S. F. (1997). Did the indo-Asian collision alone create the Tibetan plateau? *Geology, 25*(8), 719–722.

Nugteren, G., Vandenberghe, J., van Huissteden, J. K., & Zhisheng, A. (2004). A quaternary climate record based on grain size analysis from the Luochuan loess section on the central loess plateau, China. *Global and Planetary Change, 41*(3–4), 167–183.

Pan, B., Su, H., Hu, Z., Hu, X., Gao, H., Li, J., & Kirby, E. (2009). Evaluating the role of climate and tectonics during non-steady incision of the Yellow River: Evidence from a 1.24 Ma terrace record near Lanzhou, China. *Quaternary Science Reviews, 28*(27–28), 3281–3290.

Pi, K., Bieroza, M., Brouchkov, A., Chen, W., Dufour, L. J., Gongalsky, K. B., Herrmann, A. M., Krab, E. J., Landesman, C., Laverman, A. M., & Mazei, N. (2021). The cold region critical zone in transition: Responses to climate warming and land use change. *Annual Review of Environment and Resources, 46*, 111–134.

Porter, S. C., & An, Z. S. (1995). Correlation between climate events in the North Atlantic and China during the last glaciation. *Nature, 375*, 305–308.

Prell, W. L., Murray, D. W., Clemens, S. C., & Anderson, D. M. (1992). Evolution and variability of the Indian Ocean summer monsoon: Evidence from the western Arabian Sea drilling program. *Washington DC American Geophysical Union Geophysical Monograph Series, 70*, 447–469.

Pu, Y., Werne, J. P., Meyers, P. A., & Zhang, H. (2020). Organic matter geochemical signatures of sediments of Lake Ngoring (Qinghai-Tibetan plateau): A record of environmental and climatic changes in the source area of the Yellow River for the last 1500 years. *Palaeogeography, Palaeoclimatology, Palaeoecology, 551*, 109729.

Qin, Z., Storozum, M., Liu, H., Zhang, X., & Kidder, T. R. (2019). Investigating environmental changes as the driving force of agricultural intensification in the lower reaches of the Yellow River: A case study at the Sanyangzhuang site. *Quaternary International, 521*, 25–34.

Ren, M. E. (2015). Sediment discharge of the Yellow River, China: Past, present and future—A synthesis. *Acta Oceanologica Sinica, 34*(2), 1–8.

Roe, G. (2009). On the interpretation of Chinese loess as a paleoclimate indicator. *Quaternary Research, 71*(2), 150–161.

Royden, L. H., Burchfiel, B. C., & van der Hilst, R. D. (2008). The geological evolution of the Tibetan plateau. *Science, 321*(5892), 1054–1058.

Shen, J., Liu, X. Q., Wang, S. M., & Ryo, M. (2005). Palaeoclimatic changes in the Qinghai Lake area during the last 18,000 years. *Quaternary International, 136*(1), 131–140.

Shi, N. H. (1991). On the historical distribution and vicissitudes of National Vegetation of China. *Collections of Essays on Chinese Historical Geography (in Chinese), 3*, 43–73.

Shi, N. (1999). *Evolution and governance of the rivers in the Yellow River watershed.* Shaanxi People Press (in Chinese).

Shi, Y. F., Li, J. J., Li, B. Y., et al. (1999). Uplift of the Qinghai-Xizang (Tibetan) plateau and East Asia environmental change during late Cenozoic. *Acta Geography Sinica, 54*(1), 10–21.

Shi, Y., Huang, M., Yao, T., & Deng, Y. X. (2000). *Glaciers and their environments in China–the present, past and future. [Chinese.].* Science Press.

Smol, J. P., Wolfe, A. P., Birks, H. J. B., Douglas, M. S., Jones, V. J., Korhola, A., Pienitz, R., Rühland, K., Sorvari, S., Antoniades, D., & Brooks, S. J. (2005). Climate-driven regime shifts in the biological communities of arctic lakes. *Proceedings of the National Academy of Sciences, 102*(12), 4397–4402.

Song, B., Hu, F., Liang, Y., Zhang, K., Ji, J., Algeo, T. J., Hower, J. C., Yang, T., & Zhang, X. (2022). Middle to late Miocene tectonic forcing of intensified humidity in the Xunhua Basin, NE Tibetan plateau. *Palaeogeography, Palaeoclimatology, Palaeoecology, 607*, 111268.

Stevens, T., Carter, A., Watson, T. P., Vermeesch, P., Andò, S., Bird, A. F., Lu, H., Garzanti, E., Cottam, M. A., & Sevastjanova, I. (2013). Genetic linkage between the Yellow River, the mu us desert and the Chinese loess plateau. *Quaternary Science Reviews, 78*, 355–368.

Sun, X. J., Song, C. Q., Wang, F. Y., et al. (1997a). Vegetation history of the loess plateau of China during the last 100000 years based on pollen data. *Quaternary International, 37*, 25–36.

Sun, X., Song, C., Wang, F., & Sun, M. (1997b). Vegetation history of the loess plateau of China during the last 100,000 years based on pollen data. *Quaternary International, 37*, 25–36.

Sun, Y., Lu, H., & An, Z. (2006). Grain size of loess, palaeosol and red clay deposits on the Chinese loess plateau: Significance for understanding pedogenic alteration and palaeomonsoon evolution. *Palaeogeography, Palaeoclimatology, Palaeoecology, 241*(1), 129–138.

Sun, J., Xu, Q., Liu, W., Zhang, Z., Xue, L., & Zhao, P. (2014). Palynological evidence for the latest Oligocene– early Miocene paleoelevation estimate in the Lunpola Basin, Central Tibet. *Palaeogeography, Palaeoclimatology, Palaeoecology, 399*, 21–30.

Tan, Q. X. (1962). Why did the Yellow River enter a long period of stable flow after the Eastern Han: Using history to prove that rational land use in the middle course of the Yellow River was the decisive factor in eliminating floods in the lower course. *Academic Monthly, 2*, 23–55. (in Chinese).

Tang, M., & Liu, X. (1995). A new sign of delimit quaternary period. *Quaternary Research, 1995*(1), 82–88.

Tang, L., Shen, C., Lu, H., Li, C., & Ma, Q. (2021). Fifty years of quaternary palynology in the Tibetan plateau. *Science China Earth Sciences, 64*(11), 1825–1843.

Thomas, C. D. (2011). Translocation of species, climate change, and the end of trying to recreate past ecological communities. *Trends in Ecology & Evolution, 26*(5), 216–221.

Tian, L. D., Yao, T. D., Schuster, P. F., White, J. W. C., Ichiyanagi, K., Pendall, E., Pu, J., & Yu, W. (2003). Oxygen-18 concentrations in recent precipitation and ice cores on the Tibetan plateau. *Journal of Geophysical Research: Atmospheres, 108*(D9), 1.

Tian, J., Wang, P., & Cheng, X. (2004). Development of the east Asian monsoon and northern hemisphere glaciation: Oxygen isotope records from the South China Sea. *Quaternary Science Reviews, 23*(18–19), 2007–2016.

Van Breemen, N., & Buurman, P. (2002). *Soil formation*. Springer Science & Business Media.

Verosub, K. L., Fine, P., Singer, M. J., & TenPas, J. (1993). Pedogenesis and paleoclimate: Interpretation of the magnetic susceptibility record of Chinese loess-paleosol sequences. *Geology, 21*(11), 1011–1014.

Wang, S., Wu, X., Zhang, Z., Jiang, F., Xue, B., Tong, G., & Tian, G. (2002). Sedimentary records of environmental evolution in the Sanmen Lake Basin and the Yellow River running through the Sanmenxia gorge eastward into the sea. *Science in China Series D: Earth Sciences, 45*(7), 595–608.

Wang, Y., Cheng, H., Edwards, R. L., Kong, X., Shao, X., Chen, S., Wu, J., Jiang, X., Wang, X., & An, Z. (2008). Millennial-and orbital-scale changes in the east Asian monsoon over the past 224,000 years. *Nature, 451*(7182), 1090–1093.

Wang, G., Cao, K., Zhang, K., et al. (2011). Spatio-temporal framework of tectonic uplift stages of the Tibetan plateau in Cenozoic. *Science China Earth Sciences, 54*, 29–44.

Wang, X., Cui, L., Yang, S., Zhai, J., & Ding, Z. (2018). Stable carbon isotope records of black carbon on Chinese loess plateau since last glacial maximum: An evaluation on their usefulness for paleorainfall and paleovegetation reconstruction. *Palaeogeography, Palaeoclimatology, Palaeoecology, 509*, 98–104.

Wang, L., Liu, Y., Li, Q., Song, H., & Cai, Q. (2022). Tree-ring oxygen isotope recorded precipitation variations over the past two centuries in the northeast Chinese loess plateau. *International Journal of Climatology, 42*(11), 5796–5806.

Wei, K., & Gasse, F. (1999). Oxygen isotopes in lacustrine carbonates of West China revisited: Implications for post glacial changes in summer monsoon circulation. *Quaternary Science Reviews, 18*(12), 1315–1334.

Wernicke, J., Hochreuther, P., Grießinger, J., Zhu, H., Wang, L., & Bräuning, A. (2017). Air mass origin signals in δ 18O of tree-ring cellulose revealed by back-trajectory modeling at the monsoonal Tibetan plateau. *International Journal of Biometeorology, 61*(6), 1109–1124.

Wu, H., Li, X., Li, J., Jiang, Z., Li, G., & Liu, L. (2015). Evaporative enrichment of stable isotopes (δ18O and δD) in lake water and the relation to lake-level change of Lake Qinghai, northeast Tibetan plateau of China. *Journal of Arid Land, 7*(5), 623–635.

Wu, D., Ma, X., Yuan, Z., Hillman, A. L., Zhang, J., Chen, J., & Zhou, A. (2022a). Holocene hydroclimatic variations on the Tibetan plateau: An isotopic perspective. *Earth-Science Reviews, 233*, 104169.

Wu, F., Fang, X., Yang, Y., Dupont-Nivet, G., Nie, J., Fluteau, F., Zhang, T., & Han, W. (2022b). Reorganization of Asian climate in relation to Tibetan plateau uplift. *Nature Reviews Earth & Environment, 3*(10), 684–700.

Xia, D. S., Ma, Y. Z., Chen, F. H., et al. (1998). High resolution record of vegetation and climatic changes in Longxi loess plateau during Holocene. *Journal of Lanzhou University (in Chinese), 34*(1), 119-127. 40.

Xie, G. P. (1999). Pleistocene climatic and environmental change in the Lanzhou Basin, Gansu. *Acta Micropalaeontologica Sinica, 16*(1), 104–109.

Xu, Q. H., Li, M. Y., Zhang, S. R., Zhang, Y. H., Zhang, P. P., & Lu, J. Y. (2015). Modern pollen processes of China: Progress and problems (in Chinese). *Scientia Sinica Terrae, 58*, 1661–1682.

Xu, X., Zhou, B., Zhou, X., & Wang, Z. (2021). A review for the stable carbon isotope proxies in the paleovegetation researches in the loess-paleosol deposits on the Chinese loess plateau. *Quaternary Sciences, 41*(4), 931–947.

Yang, S. Y., Cai, J. G., Li, C. X., & Deng, B. (2001). New discussion about the run-through time of the Yellow River. *Marine Geology and Quaternary Geology, 21*(2), 17–22.

Yang, J., Yuan, H., Hu, Y., & Wang, F. (2022). Significance of sedimentary provenance reconstruction based on borehole records of the North China plain for the evolution of the Yellow River. *Geomorphology, 401*, 108077.

Yi, S., Saito, Y., Oshima, H., Zhou, Y., & Wei, H. (2003). Holocene environmental history inferred from pollen assemblages in the Huanghe (Yellow River) delta, China: Climatic change and human impact. *Quaternary Science Reviews, 22*(5–7), 609–628.

Yuan, D., Cheng, H., Edwards, R. L., Dykoski, C. A., Kelly, M. J., Zhang, M., Qing, J., Lin, Y., Wang, Y., Wu, J., & Dorale, J. A. (2004). Timing, duration, and transitions of the last interglacial Asian monsoon. *Science, 304*(5670), 575–578.

Zachos, J., Pagani, M., Sloan, L., Thomas, E., & Billups, K. (2001). Trends, rhythms, and aberrations in global climate 65 Ma to present. *Science, 292*(5517), 686–693.

Zhang, J., Harbottle, G., Wang, C., & Kong, Z. (1999). Oldest playable musical instruments found at Jiahu early Neolithic site in China. *Nature, 401*(6751), 366–368.

Zhang, E., Shen, J., Wang, S., Yin, Y., Zhu, Y., & Xia, W. (2004). Quantitative reconstruction of the paleosalinity at Qinghai Lake in the past 900 years. *Chinese Science Bulletin, 49*(7), 730–734.

Zhang, H., Yang, M., Zhang, W., Lei, G., Chang, F., Pu, Y., & Fan, H. (2008). Molecular fossil and paleovegetation records of paleosol S4 and adjacent loess layers in the Luochuan loess section, NW China. *Science in China Series D: Earth Sciences, 51*(3), 321–330.

Zhang, J. F., Wang, X. Q., Qiu, W. L., Shelach, G., Hu, G., Fu, X., Zhuang, M. G., & Zhou, L. P. (2011). The paleolithic site of Longwangchan in the middle Yellow River, China: Chronology, paleoenvironment and implications. *Journal of Archaeological Science, 38*(7), 1537–1550.

Zhang, J., Lu, H., Gu, W., Wu, N., Zhou, K., Hu, Y., Xin, Y., & Wang, C. (2012). Early mixed farming of millet and rice 7800 years ago in the middle Yellow River region, China. *PloS One, 7*(12), e52146.

Zhang, S. B., Zheng, Y. F., Wu, P., He, Q., Rong, W., Fu, B., Yang, Y. H., & Liang, T. (2020). The nature of subduction system in the Neoarchean: Magmatic records from the northern Yangtze craton, South China. *Precambrian Research, 347*, 105834.

Zhang, Y., Zhang, J., McGowan, S., Metcalfe, S., Jones, M., Leng, M. J., & Hou, J. (2021). Climatic and environmental change in the western Tibetan plateau during the Holocene, recorded by lake sediments from Aweng co. *Quaternary Science Reviews, 259*, 106889.

Zhang, S., Sun, Y., Li, M., Wang, N., & Xu, Q. (2022). Paleovegetation and paleotemperature in North China during the mid-Holocene based on sedimentological and palynological evidence from Lake Baiyangdian. *Palaeogeography, Palaeoclimatology, Palaeoecology, 595*, 110982.

Zhao, W., Chen, C., Jiang, Q., Ji, M., Zhen, J., Yang, Y., Ning, D., Li, X., & Shen, J. (2021). Holocene hydroclimate in the source region of the Yellow River: A new sediment record from Lake Gyaring, NE Tibetan plateau. *Journal of Asian Earth Sciences, 205*, 104601.

Zheng, B., Xu, Q., & Shen, Y. (2002). The relationship between climate change and Quaternary glacial cycles on the Qinghai–Tibetan Plateau: review and speculation. *Quaternary International, 97*, 93–101.
Zheng, J., Wang, W. C., Ge, Q., Man, Z., & Zhang, P. (2006). Precipitation variability and extreme events in eastern China during the past 1500 years. *TAO. Terrestrial, Atmospheric and Oceanic Sciences, 17*(3), 579.
Zhuang, Y., & Kidder, T. R. (2014). Archaeology of the Anthropocene in the Yellow River region, China, 8000–2000 cal. BP. *The Holocene, 24*, 1602.

Chapter 4
Soil Erosion: Controlling Factors and Processes

Contents

4.1 Introduction

Soil erosion in the Yellow River basin is well known throughout the world. The middle reaches of the Yellow River, the Loess Plateau, contribute over 80% of the total sediment in the Yellow River. Before the 1980s, an area of 472,000 km^2 in the Loess Plateau experienced significant soil and water loss (Zhao et al., 2013); An annual average of 1.6 billion tons of sediment entering the Yellow River, among which 0.4 billion tons of sediment deposited in the lower Yellow River channel (Gong, 1987). Such sedimentation created an elevated river, which means the riverbed is higher than the surrounding flood plan, therefore, it poses great flood risks. Severe soil erosion stripped away surface soil and organic materials. It was

T. Zhu, *Watershed Environmental Degradation and Sustainable Development*,
https://doi.org/10.1007/978-3-032-02448-0_4

estimated that an average of 1 cm of soil was removed by erosion each year in the Loess Plateau (Tsunekawa et al., 2014). The removal of surface soils and nutrients leads to a decrease in crop productivity. The sediments and nutrients entering the water body also cause sedimentation and water contamination. Gully development dissects the terrain and reduces arable land in the region. Soil erosion is considered as the root cause of economic development and poverty alleviation in the region. Soil loss has significantly declined since the 1980s, particularly after 2000, largely due to extensive soil conservation efforts in the region. On the Loess Plateau, the sediment load decreased by 74% in the period of 2008 to 2016, compared to the period of 1971 to 1987, with soil conservation efforts responsible for 72% of this reduction (Zhang et al., 2019). As a result, the annual sediment load in the Yellow River was significantly decreased, averaging 0.25 billion tons per year between 2000 and 2018 (Dang et al., 2020). Soil erosion is the primary environmental concern in the Yellow River basin, and controlling it is essential for sustainable development. This chapter will examine the factors that influence soil erosion and the major erosion processes on the Loess Plateau. The following chapter will discuss soil conservation strategies for the same region.

4.2 Controlling Factors of Soil Erosion

4.2.1 Rainfall

The Loess Plateau is a typical continental monsoon climate, characterized by dry, cold winters and warm, wet summers. Annual precipitation varies from approximately 200 mm in the northwest to 800 mm in the southeast, with an average of around 440 mm (Feng et al., 2016). Over 70% of the annual precipitation falls during the summer months, from June to September, and most runoff-generating storms occur during this period. We analyzed the runoff-generating rainfall events in the Wangjiagou watershed of Shanxi Province and found that the average event rainfall was 21.9 mm, ranging from 3.6 mm to 110 mm. The average rainfall duration was 3.4 h, with a range of 0.25 h to 26.1 h, and the average intensity was 10.3 mm/h, ranging from 1.03 mm/h to 62.4 mm/h (Zhu & Zhu, 2014). A range of rainfall indexes have been used to quantify the rainfall erosivity, such as the event mean rainfall intensity (Zhou & Wang, 1987; Wei et al., 2014), the maximum 30-minute rainfall intensity (Zheng, 2006; Li et al., 2022a, 2022b; Han et al., 2017), effective rainfall amount (Zhu & Xu, 2021), rainfall duration, and rainfall patterns (Wei et al., 2007; Fang et al., 2008; Han et al., 2017).

Soil erosion is also affected by the temporal changes of precipitation in the region. Sun et al. (2015) analyzed data from 299 weather stations on the Loess Plateau to examine changes in precipitation between 1961 and 2011, revealing a general decrease in annual precipitation across most of the region. Similarly, Xin et al. (2011) reported that annual precipitation, erosive precipitation (defined as daily precipitation >12 mm), and rainfall erosivity declined by 9.4%, 11.0%, and 13.5%, respectively, from the period 1956–1979 to 1980–2008. In their study, annual erosivity was calculated by summing the EI30 index values for all erosive events.

Although extreme rainfall events may constitute a small proportion of the total precipitation, they can lead to significant soil loss. Zhu (2016) studied the long-term soil loss from two neighboring small watersheds in the Loess Plateau and found that rainfall events with return periods of 2 years or more accounted for only 12.8% of the total precipitation but contributed to 70.2% and 75.4% of the soil loss in the two watersheds, respectively. Li et al. (2022a, 2022b) attributed the increase in soil erosion rates between 2010 and 2016, following a decline from 1980 to 2000, to the rising frequency of extreme rainfall events. However, Li et al. (2010) compared precipitation data from 1960–1990 and 1990–2007, finding no significant changes in extreme precipitation events in the latter period.

4.2.2 Soil Properties

The middle reach of the Yellow River is blanketed by loess ranging from 0 to 400 m in thickness (Yu et al., 2020). Spatially, the wind-blown loess can be classified into sandy loess belt, loess belt and clayey loess belt from northwest to southeast. The particle size of sand fraction of the surface loess (>0.075 mm) decreases from 45% in the northwest to 6% in the southeast of the region whereas the clay content increases from 9% to 31% (Li et al., 2022a, 2022b); However, the mineralogy of the coarse silt, fine silt and clay fractions of the uppermost loess layer is generally uniform across the Loess Plateau (Eden et al., 1994). Vertically, the Quaternary loess can be divided into three stratigraphical units from oldest to youngest: Wucheng, Lishi and Malan Loess that deposited in the early Pleistocene (Q1), Middle Pleistocene (Q2), and late Pleistocene period, respectively (Liu, 1965). The loess is often underlaid by the Tertiary clayey red earth in the region. The red earth consists of an average of 13.3% sand (>0.05 mm), 38.1% silt (0.05–0.005 mm) and 56.5% of clay (<0.005 mm) (Zhu, 2003). Additionally, Pisha sandstone—a thick layer of loosely bound, fluvial-clastic sandstone, shale, and mudstone that rapidly disintegrates upon contact with water—is widely distributed in the northern part of the region (Liang et al., 2019).

The loess soils have a bulk density ranging from 1.10 to 1.40 g/cm^3. They are high in silt content and low in organic matter, becoming highly unstable when exposed to water. Water stable aggregates content is generally less than 5% (Luk et al., 1989). While the original loess deposits usually have high shear strengths (> 100 kPa), the cultivated loess soils are relatively low in soil shear strengths, generally ranging from 5–20 kPa (Zhu et al., 2021). Therefore, loess soils have high erodibility in water erosion.

Since surface erosion is caused by infiltration-excess overland flow, the infiltration capacity of soils in the region plays a vital role in runoff generation and soil erosion. The stable infiltration capacity varied with vegetation cover, land use, and soil properties with a range of 0.42–1.11 mm/min (Li et al., 1992). Loess soil is also prone to the formation of surface crusts that can greatly reduce the infiltration rates and thereby increase runoff generation and soil loss during storm events. Rainfall simulated experiments showed that surface crusts could reduce stable infiltration rate from 0.8 mm/min to 0.3 mm/min (Cai et al., 1990). In the cultivated land, plough-pan caused by compaction of plough also decreases infiltration rates within the soil profile, which could significantly increase runoff generation during large storm events when rainwater accumulates above the plough pan.

4.2.3 Topography

Topography in the Loess Plateau is characterized by inter-gully and gully areas, separated by a sharp slope break, or gully edge line (Fig. 4.1). The inter-gully area features gentle or medium slope gradients ranging from a few degrees near the catchment divides to more than 20 degrees towards the gully edge. Soil erosion in the inter-gully area is dominated by sheet erosion and rill erosion, with occasional ephemeral gully erosion and soil piping. Gully areas are characterized by a drastic increase in slope gradient, typically greater than 30 degrees and up to 90 degrees.

Fig. 4.1 Gully and inter-gully areas on the Loess Plateau

Soil erosion is dominated by mass movement, tunnel erosion and gully/channel erosion.

The effect of topography on soil erosion has been extensively investigated in the region. Such studies can be categorized by field investigations and experimental studies. Numerous field erosion plots with different slope gradients, slope length have been set up in different locations across the region and monitored under natural rainfall conditions (Liu et al., 1994; Liu et al., 2000; Sun et al., 2014; Fang et al., 2008; Qin et al., 2018; Ma et al., 2019). In addition to the field studies, experimental research has been carried out under simulated rainfall conditions (Luk & Cai, 1990; Luk et al., 1993; Fang et al., 2015; Shen et al., 2015; Shen et al., 2016). Empirical regression models have frequently been developed to establish correlations between soil loss and topographic parameters (e.g. slope gradient, slope length etc.). Research in the region indicates that soil erosion increases with slope gradient, following either a power function (Wu et al., 2017), linear function (Wang et al., 2018), or binary function (Guo et al., 2021). One of the notable contributions from those studies is to examine the effect of slope gradient on soil loss on the steep slopes which are not incorporated in the USLE model (Liu et al., 1994, 2000). The relationship between soil erosion and slope length is more complicated. On the one hand, with the increase of slope length, more runoff generated, which can increase the erosive power and thereby erosion (He et al., 2022). On the other hand, with more sediment carried by flow and more energy will be consumed and therefore, weakening erosion. As a result, the relation between slope length and erosion is not a simple linear relation. Research on the effect of slope shape on erosion in the region has been limited. Slope shapes are generally classified into three types: concave, convex, and straight. The region is predominantly characterized by convex slopes, where the upper slope is gentler, and the lower slope is steeper, with the two sections separated by the gully edge line.

4.2.4 Land Use and Land Cover

Land use on the Loess Plateau is currently dominated by small land parcels managed by individual households. The main land use types include terrace cropland, slope cropland, grassland, shrubland, forestland, barren or sparsely vegetated valley slopes, and check-dam land. Long-term monitoring in the Wangjiagou experimental watershed showed that soil loss decreases in the following order: slope cropland < shrubland < forestland < grassland < terrace land. Compared to cropland, soil loss was reduced by 74.9% on shrubland, 89.4% on forestland, 93.1% on grassland, and 93.6% on terrace land (Zhu & Zhu, 2014). Wei et al. (2014) found that the mean runoff coefficient and erosion modulus followed this order: cropland > pastureland > woodland > grassland > shrubland. The differences between these studies may be due to variations in plant age and species, slope conditions, observation duration, and rainfall patterns. While soil loss is typically minimal on check-dam lands, it can be significant during extreme storm events if the dams are breached.

Land cover change is the primary human-driven factor contributing to increased soil loss in the region. Intensive agricultural activities have cleared natural vegetation, converting the land for farming purposes. A number of studies have been carried out on the impact of vegetation on soil loss in the Loess Plateau region. Zheng (2006) found that the destruction of vegetation could increase soil loss by 797 to 1697 times compared to the forested land before the damage. She also noted that vegetation plays a far more significant role in soil loss than rainfall and topography. Chen et al. (2018) observed that grassland (Bothriochloa ischaemum L.) and shrubland (Sea-buckthorn) with high ground cover were more effective at reducing soil loss than planted forests with poor ground cover. These areas had lower runoff depth and soil loss compared to forestlands with insufficient ground cover. Similarly, Wang et al. (2016) found that grasslands and woodlands were more effective at reducing erosion than orchards. The "Grain for Green" Program, launched by the Chinese government in 1999, has significantly altered land cover in the region by converting steep croplands and barren lands into forests, grasslands, and shrublands, leading to a reduction in soil loss. Numerous reports have documented a dramatic decrease in soil loss in the region by comparing land use changes before and after the program's implementation. Yu et al. (2020) reported that from 2000 to 2010, farmland decreased by 28.3%, while grassland and forest coverage increased by 12.4% and 5.0%, respectively. Fu et al. (2011) used the Universal Soil Loss Equation (USLE) to evaluate the program's impact between 2000 and 2008, finding that the average soil retention rate (the percentage of soil retention relative to potential soil loss) reached 63.3%. During this period, soil loss rates decreased in 34% of the Loess Plateau, remained unchanged in 48%, and slightly increased in 18% of the region.

4.2.5 Soil Tillage

On cropland, traditional tillage practices in the region involve shallow plowing and the removal of crop residues. In contrast, no-tillage practices with crop residue cover can enhance soil quality, minimize surface disturbance, and significantly reduce soil loss (Zhang et al., 2004; Wang et al., 2008a, 2008b; Yu et al., 2020). A 16-year field experiment in Shanxi demonstrated that long-term no-tillage with straw cover increased soil organic matter (SOM) by 21.7%, total nitrogen (TN) by 51.0% at a 0–10 cm depth, and available phosphorus (P) by 97.3% at a 0–5 cm depth compared to traditional tillage (Wang et al., 2008a, 2008b). The study also reported that no-tillage with residue cover can reduce water and soil loss by approximately 40% and 80%, respectively, on moderately sloping farmland.

4.3 Soil erosion Processes

Soil erosion processes in the region include rainsplash erosion, sheet erosion, rill erosion, gully erosion, mass movement, tunnel erosion as well as wind erosion. Spatial patterns of those erosion exist at both watershed level and regional scale.

4.3.1 Rain Splash, Sheet, and Rill Erosion

Rain splash not only causes the downslope movement of soil particles but also breaks down soil aggregates into fine particles, making them more easily transported by sheet and rill flows. Loess soils are particularly vulnerable to rain splash erosion due to their low aggregate stability and high silt content. Rainfall simulation experiments have been conducted to examine how factors like rainfall intensity, soil properties, slope gradient, soil crust formation, and antecedent soil moisture affect rain splash on loess soils. Typically, soil loss from rain splash increases with both rainfall intensity and slope gradient, although soil detachment is more sensitive to rainfall intensity (Wu et al., 2019a, 2019b). The amount of splash detachment and raindrop size also increases in a positive linear relationship (Fu et al., 2016). Splash rates are highest in sandy loam from the northern Loess Plateau and lowest in clay loam from the southern part, following a decreasing trend: sandy loam > sandy clay loam > clay loam > loamy clay (Sun et al., 2021). Large raindrops are particularly responsible for soil detachment. The breakdown of micro-aggregates by rain splash leads to soil surface sealing and clogging of soil pores, which reduces infiltration capacity and increases surface runoff. However, Luk and Cai et al. (1990) found that soil crust formation on loess surfaces can increase shear strength by up to 11 times and reduce splash soil loss by as much as 76%.

Sheet erosion is directly influenced by factors such as slope gradient, rainfall intensity, tillage practices, and the sediment concentration in sheet flows (Fig. 4.2). These relationships have been studied and quantified through rainfall simulation experiments in the region. Sheet erosion tends to increase either linearly or exponentially with slope gradient and rainfall intensity, but it decreases as a power function as the sediment load in sheet flow increases on bare cropland (Zhang et al., 2019; Wu et al., 2019a, 2019b). Reducing tillage can significantly lower sheet erosion rates. In rangelands, sheet erosion increases with both rainfall intensity and slope gradient but decreases with higher vegetation cover, with stream power being the most reliable hydraulic predictor of sheet erosion rates (Wang et al., 2018).

Rill erosion is considered a major issue, accounting for approximately 70% of erosion on sloped farmlands in the Loess Plateau (Zheng & Tang, 1997). The formation and progression of rills involve several processes, including the development of nickpoints, downcutting by rill flows, lateral erosion from soil collapse, headward erosion, and soil piping (Cai, 1998; Sun et al., 2013). Rill erosion is influenced by

Fig. 4.2 Sheet erosion on the slope cropland in the Wangjiagou watershed of Shanxi Province

topographic factors like slope gradient and slope length (Zheng, 1989; Wang, 1998; Chen et al., 2016), rainfall and runoff (Zhu et al., 1997; Shi et al., 2012a; Xu et al., 2022), soil properties (Shen et al., 2021), and surface cover (Shi et al., 2012b). The rate of rill erosion increases with rainfall intensity, slope gradient, and slope length. While sheet flow erosion in interrill areas tends to selectively remove fine particles, rill flow is less selective because of its higher discharge rate and greater flow velocity compared to sheet flow (Shi et al., 2012b).

4.3.2 Gully Erosion

Gully erosion refers to the erosion of channels formed by concentrated water flow, creating channels too large to be removed by standard tillage practices (FAO, 1965). Gully erosion is highly active on the Loess Plateau (Fig. 4.3) and is a major source of sediment production, accounting for approximately 60–90% of the total sediment yield in the area (Liu et al., 2012). The gullies in the region are categorized into four types: headwater gullies, hillside gullies, valleyside gullies, and valley-floor gullies (Zhu, 2016). A headwater gully forms in the topographically convergent zone of a valley or stream's headwater area. Multiple headwater gullies often merge downstream to form larger valleys or ephemeral stream channels. Hillside gullies develop

Fig. 4.3 Gully development on the Loess Plateau

on hillslopes and join perpendicularly to larger valleys or ephemeral stream channels. Valleyside gullies form on the sideslopes of valleys, while valley-floor gullies cut into alluvium on broad valley bottoms along the channel flow path.

The gully erosion rate is a crucial parameter for assessing the impact of gully formation on sediment production and for quantifying the relationship between gully erosion and its controlling factors. Various methods have been used in the region to measure gully erosion rates. Wu and Cheng (2005) used GPS to measure the morphology of gullies in a small catchment and reported that the retreat rate of hillside gullies ranged from 0.16 to 2.02 m per year. Li et al. (2017a, 2017b) combined a 3D scanner with QuickBird images to estimate gully volume changes in two small catchments on the central plateau, with values ranging from 0.41 to 36.76 m^3 per year, and an average of 10.32 m^3 per year. Guan et al. (2021) utilized a combination of methods, including unmanned aerial vehicles (UAV), historical aerial photos, and field surveys, to assess long-term gully development rates (1959–2018) in the northeastern Loess Plateau. They found that the eroded volume from gully-head retreat was 6.1 times greater than that from gully sidewall retreat, indicating that gully-head retreat was the primary driver of gully development during the study period. Wang et al. (2021) evaluated gully development rates in the tablelands of the southeastern Loess Plateau using KH-4B satellite images, Quickbird-2 images, and UAV images. They found that the rates (RA) for 45 expanding gullies ranged from 0.50 to 20.94 m^2 per year, with an average of 5.66 m^2 per year between 1969 and 2010. In all of these studies, an empirical power function model was developed to link gully development rates to topographic parameters, such as drainage area and slope gradient, within the gully drainage areas.

The development of dense gullies in the region is a result of both natural and human factors. Loess soils are widely recognized for their high susceptibility to gully formation and development globally (Thomas & Welch, 1988; Nachtergaele et al., 2001; Valentin et al., 2005; Nachtergaele & Poesen, 2002; Kertész & Gergely, 2011; Zhu, 2012; Kariminejad et al., 2020). Gully formation typically requires a critical drainage area (A) and a corresponding drainage slope gradient (S) to generate sufficient concentrated flows. The critical S–A topographic threshold relationship ($S = aA^b$) has been established to predict potential gully initiation locations (Wu & Cheng, 2005; Guan et al., 2021). Tang et al. (1998) found that the critical slope gradient for gully initiation on sloped croplands ranges from 15° to 20°, with an average of 18.1°. The coefficients a and b vary depending on factors such as soil type, vegetation cover, and land use. Soil sealing on loess surfaces reduces infiltration, increasing surface flow and promoting gully formation. Road construction can also accelerate gully development. Conversely, terracing sloped land and planting vegetation can help stabilize and prevent further gully growth.

Gully development is closely linked to other erosion processes, such as mass movement and tunnel erosion in the region. Mass movement contributes to gully expansion by advancing the gully head upslope/upstream and widening the gully. It also decreases soil shear strength, making the soil more susceptible to erosion by concentrated gully flows. Tunnel erosion can occur above the gully head, along the gully channel, or on the gully sideslopes. The formation of tunnels can cause gully heads to move upslope, extend the gully network, and deepen the gully channel (Zhu, 2012; Zhu et al., 2020). In a small watershed, Wangjiagou, Shanxi Province on the eastern Loess Plateau, tunnels were found to be associated with 79% of headwater gullies, 48% of hillside gullies, 3% of valleyside gullies, and none of the ephemeral river channels (Zhu, 2012).

4.3.3 Mass Movement

Mass movement is highly active in the Loess Plateau and poses not only a major natural hazard, causing casualties and property damage (Derbyshire et al., 1994; Wang et al., 2011; Xu et al., 2017), but also significantly contributes to sediment production (Zhu et al., 2019). The primary types of mass movement in the region include slides (Fig. 4.4), earth falls (Fig. 4.5), mudflows, and soil creep.

Various factors influence the occurrence of mass movements. Loess soils are particularly susceptible to these movements. Shallow slides are commonly associated with Malan Loess, which has a high silt content, while earth falls are frequent in Lishi Loess, where vertical cracks are densely distributed. Soil creep is most pronounced in the Tertiary red earth, often triggered by freeze-thaw or wet-dry cycles due to its high clay content. Spatially, earth falls tend to occur along gully edges, slides can occur both above and below the gully edge, and soil creep is found on valley sideslopes where red earth is exposed. Mudslides typically happen along gully or valley channels. Topographic thresholds for mass movement vary

Fig. 4.4 Shallow slides on the Loess Plateau

depending on the type. Slides typically occur on slopes between 40 and 50 degrees, while falls are more common on slopes steeper than 50 degrees. Guo et al. (2019a, 2019b) conducted rainfall simulation experiments on natural loess slopes and found that shallow mass movement rates increased by approximately 3–8 times when slope height increased from 1.0 m to 1.5 m, and by 148%–525% as the slope gradient rose from 60° to 70°. In terms of land use, most mass movements occur on barren or sparsely vegetated slopes. However, Guo et al. (2020) also found that vegetation cover can increase the likelihood of shallow landslides in the region.

Studies on sediment production by mass movement in the region are relatively limited. Zhu and Xu (2021) conducted a survey of all active mass movements in the Wangjiagou watershed with a drainage area of 9.1 km^2 in Shanxi and found that the majority of mass movements (88.6%) had volumes between 1 and 10 m^3, but these events contributed only 30.6% of the total mass movement volume in the watershed. In contrast, mass movements with volumes exceeding 1000 m^3 accounted for just 0.2% of all events, yet contributed 22.6% of the total volume. Despite the focus typically being on large mass movements, small and medium-sized events (less than 1000 m^3) contribute the majority of sediment generated by mass movement (77.4%).

Mass movements generally do not directly deliver sediment to the rivers except they occur along the river banks. But mass movement debris is much more prone to water erosion due to the drastic decrease in soil shear strength. In the same above-mentioned small watershed, Zhu et al. (2020) compared the shear strength of loess

Fig. 4.5 Loess falls on the Loess Plateau

and red earth before and after mass movement, it was found that freezing and thawing actions on the Tertiary red earth produce loose soil debris, which virtually decreases the red earth materials to nearly 0 kPa of soil shear strength. Slab falls decrease the shear strength of the original Lishi Loess from over 100 kPa to a mean shear strength of about 9 kPa. Shallow slides also reduce the undisturbed slope materials to an average shear strength of 6.3 kPa.

Regarding the contribution of sediment from mass movements, Zhu (1989) reported that the total volume of mass movement in the Wangjiagou watershed was about 25% of the sediment yield. However, this ratio does not represent the direct contribution of mass movement to sediment yield because of the sediment delivery ratio. Typically, only a small portion of mass movement debris is transported by water flow to the watershed outlet. Measuring this delivery ratio under natural conditions is challenging. Nevertheless, rainfall simulation experiments can help monitor mass movement occurrences and evaluate their role in sediment production (Xu et al., 2015a, 2015b, 2015c), though the issue of scale remains. In other words, the relatively small mass movements observed in simulated rainfall experiments may not be comparable to the larger-scale mass movements that occur under natural conditions with regard to sediment delivery ratio.

4.3.4 Tunnel Erosion

Tunnels developing in the Loess Plateau are among the largest size in the non-karst regions in the world, up to 20 m or more in both depth and diameter (Zhu, 1997). The density and size of tunnels generally decreases from northwest to southeast in the region (Wang et al., 2024). The formation and development of tunnels are influenced by a range of factors, including the thickness and particle size of loess deposits, the presence of joints and fissures in the loess, variations in particle size along the vertical profile, surface topography, vegetation cover and root systems, as well as climatic conditions, land use. among other factors (Li et al., 2019; Wang et al., 2020, 2024). Tunnel formation seems to be a mechanical-dominant process by surface flow entering underground. When vertical movement of water encounters a relatively low permeable layer, water excavates downslope above the layer and develops subterrain tunnels. A typical tunnel consists of tunnel inlet (Fig. 4.6), tunnel path (Fig. 4.7) and tunnel outlet (Fig. 4.8). Most tunnels are formed in few catastrophic storm events. The size of tunnels developed in the Malan loess is generally smaller than those developed in Lishi loess whereas there is no tunnel development in the Tertiary red earth. Majority of tunnels are developed below the gully edge line, though tunnels are also found on the slope cultivated land and terrace land. Tunnel enlargement can occur in storm or inter-storm periods through topples of

Fig. 4.6 Tunnel inlet

Fig. 4.7 Tunnel path

inlet walls, and mechanical erosion and roof cave-in collapses in tunnel paths (Zhu, 2003).

Tunnel flows occur only after the onset of rainfall, following a brief delay after surface flow begins. Although tunnel flow discharges originate from surface drainage areas through tunnel inlets, they do not exhibit a strong correlation with rainfall

Fig. 4.8 Tunnel outlet

parameters, unlike overland flow. This is attributed to frequent blockages within the tunnels due to collapses, which may reopen during subsequent events, as well as the occasional sudden formation of new inlets (Zhu, 1997). Additionally, there are generally no significant correlations between runoff and sediment yield in most tunnels, both within individual storms and between different storm events (Zhu et al., 2002).

The significance of tunnel flows in contribution of catchment outflow and sediment yield was not well understood due to lack of field studies in the region. One exception is that a two-year monitoring program conducted in a small headwater catchment in the hilly-gully loess region. The results show that at least 57% of the catchment sediment and 43% of the catchment flow discharges were routed via tunnel systems (Zhu, 1997; Zhu et al., 2002). The net erosion of tunnels may contribute to 25–30% of the catchment sediment yield.

4.3.5 Wind Erosion

Wind erosion is a significant process in the wind-water erosion crisscross region of the Loess Plateau (Fu, 1989). It is driven by factors such as readily erodible sandy loess, sparse vegetation cover, strong winds, overgrazing, and overcultivation. It primarily occurs in the winter and spring when the vegetation cover is lacking. The loose, redeposited wind-blown soil particles become highly vulnerable to water erosion during the summer months. However, research on wind erosion is less extensive than on water erosion due to the challenges of studying it with limited methods, such as soil flux samplers (Guo et al., 2019a, 2019b; Tuo et al., 2018). The estimated contribution of wind erosion to total erosion varies significantly among different studies. Guo et al. (2019a, 2019b) measured both wind and water erosion on the agricultural cropland plots and found that annual water erosion ranged from 8.1 to 89.4 t ha − 1 a − 1, with an average of 43.8 t ha − 1 a − 1, while wind erosion ranged from 0.82 to 2.12 t ha − 1 a − 1, with an average of 1.58 t ha − 1 a − 1. This indicates that wind erosion accounted for only 3% of the total erosion. In contrast, Tuo et al. (2018) reported that wind erosion, estimated at 1460 t km − 2 a − 1 on northwest-facing slopes, contributed approximately 39.7% of total soil loss by subtracting water erosion (calculated using the USLE model) from total erosion (estimated through cesium-137 inventories). Similarly, Zhang et al. (2018) found that wind erosion contributed an average of 27.4% to total erosion on windward slopes (northwest and north), with deposition occurring on the opposite leeward slopes (southeast and south). Clearly, a systematic and long-term study is necessary to fully evaluate the overall impact of wind erosion in the region.

4.4 Modeling of Erosion Processes and Sediment Transport

Numerous studies have focused on simulating runoff, erosion, and sediment transport processes in the Yellow River Basin, particularly in the middle reaches—the Loess Plateau (Hessel et al., 2003a, 2003b; Hessel & Van Asch, 2003; Wang et al., 2007; Yang et al., 2012). These studies commonly apply process-based erosion models such as WEPP and LISEM, which are typically used at the small watershed scale. Watersheds are often subdivided into intergully and gully areas: in intergully areas, erosion is primarily driven by interrill and rill processes, whereas gully areas are dominated by gully erosion, tunnel erosion, and mass movement. Runoff is primarily generated through infiltration-excess (Hortonian) flow due to the semi-arid climate. Infiltration was modeled using a modified Green-Ampt infiltration equation.

$$f(t) = Ks\left(\frac{\psi^{*}\Delta\theta}{F(t)} \cdot +1\right) \tag{4.1}$$

where f(t) = infiltration rate (cm/h); Ks = saturated hydraulic conductivity (cm/h); ψ = matric suction head (cm); $\Delta\theta$ = change in volumetric moisture content; F(t) = cumulative infiltration depth at time t.

Interrill and rill erosion are calculated separately using the equations (Eqs. 4.2 and 4.3) developed in the WEPP model (Nearing et al., 1989; Laflen et al., 1991). These equations have been applied in studies of the Loess Plateau region (Han et al., 2016; Zheng et al., 2020; Zhu et al., 2023).

Interill erosion:

$$E_i = K_i \cdot I_e^2 \cdot SDR \cdot C_f \tag{4.2}$$

where E_i = interrill erosion rate ($kg \cdot m^{-2} \cdot s^{-1}$); K_i = interrill erodibility ($kg \cdot s \cdot m^{-4}$); I_e = effective rainfall intensity ($m \cdot s^{-1}$); SDR = sediment delivery ratio; C_f = cover and management factor.

The LISEM model includes a similar equation for interrill erosion but uses rainfall kinetic energy instead of effective rainfall intensity. In both the WEPP and LISEM models, rainsplash erosion is incorporated into interrill erosion through the use of either rainfall intensity or kinetic energy.

With respect to rill erosion, the following equation was used in WEPP.

$$Dr = Kr(\tau_f - \tau_c)\left(1 - \frac{Tc}{Sa}\right) \tag{4.3}$$

where Dr. = rill detachment rate ($kg/m^2/s$); Kr = rill erodibility parameter (s^2/m^2); τ_f = hydraulic shear stress exerted by flow in the rill (N/m^2); τ_c = critical shear stress for soil detachment (N/m^2); S_a = actual sediment load in the rill flow (kg/m/s); T_c = sediment transport capacity of the flow (kg/m/s).

In gully areas, the modeling of erosion processes remains limited. For example, Hessel et al. (2003a) developed an equation to estimate sediment deposition resulting from slab failures within gully systems.

$$M = Y_d Hd(0.5DX) \tag{4.4}$$

where M = mass of soil fall (kg); Yd = dry bulk density (kg/m3); H = height of failed slab (m); d = slab thickness (m); and DX = pixel size (m). However, mass movement has other forms, such as slides, mudflows, and soil creep which have not modelled in the region for sediment production.

Zhu et al. (1992) developed tunnel erosion using the field observation data in the YDG catchment.

$$S = e^{5.13} Q_h^{\ 1.04} H^{0.071} L^{0.13} \tag{4.5}$$

where: S = sediment load (kg); Q_{h} = Runoff depth (mm); H = Elevation difference between tunnel inlet and outlet (m); L = length of tunnel (m).

Sediment transport within gully systems is also constrained by the sediment delivery ratio. An empirical model developed by Luk et al. (1992) was used to estimate the proportion of sediment that is actually delivered through gully systems.

$$R_{sd} = 0.0277 P^{-0.29} C^{0.19} S_m^{\ 0.59} (E_a / E)^{0.44} \tag{4.6}$$

where R_{sd} = sediment delivery ratio;P = rainfall amount (mm); C = runoff coefficient; E_a/E = proportion of kinetic energy with rainfall intensity >0.15 mm.

Once sediment is routed into river systems, deposition may occur when the sediment load exceeds the river's transport capacity. However, in cases of high sediment concentration, the transport capacity of the river flow differs from that of low-sediment flows (Cheng et al., 2024). To account for this, Zhang and Zhang (1992) developed a sediment transport capacity equation specifically for high sediment-laden flows, as follows:

$$S_* = K \left[\frac{(0.0022 + S_v) U^3}{\kappa \dfrac{\rho_s - \rho_m}{\rho_m} gh\omega_m} \ln\left(\frac{h}{6D_{50}} \right) \right]^M \tag{4.7}$$

where κ = von Karman coefficient in muddy water; U = flow velocity(m/s); h = depth (m) *Sv* = volumetric concentration for the total sediment; ω_m = average settling velocity of non-uniform suspended load and *D50* = median diameter of bed material, K, M = coefficients.

4.5 Limitations of the Previous Research and Recommendations for Future Studies

The Loess Plateau is one of the most severely eroded regions in the world, susceptible to all major types of erosion, including sheet erosion, gully erosion, mass movement, tunnel erosion, and wind erosion. With more control efforts being implemented in inter-gully areas, erosion from gully regions has become increasingly significant. While there is a wealth of research on sheet erosion, fewer studies focus on gully erosion and mass movement, and even fewer address tunnel erosion and wind erosion.

Most current field studies are short-term, and their results typically reflect soil loss under small to medium rainfall events as the extreme rainstorms are rare. Therefore, soil erosion under extreme rainfall conditions requires further investigation. Although extreme storms make up a small percentage of total rainfall in a region, they generate a disproportionately large amount of sediment (Zhu, 2016). Furthermore, other types of erosion, such as mass movement, gully erosion, and tunnel erosion, may become highly active during extreme rainstorms. As a result, the relative importance of different types of erosion observed in small to medium storms may differ significantly from that during extreme storm events. Soil erosion under storms of varying frequency and magnitude needs to be further studied in the future.

Current field studies on soil erosion have often been conducted on small, closed-boundary erosion plots where sheet erosion is dominant. The erosion rates observed within these closed plots are then extrapolated to similar surface conditions within the watershed, which can lead to significant errors. Since runoff generated from upslope areas outside the plot is blocked by the plot boundaries, its impact on erosion within the plot is often overlooked. Similarly, the flow generated within the plot, which causes downslope erosion, is not accounted for in the plot's erosion measurements. In the future, more open, hierarchical plot/catchment field monitoring studies should be conducted.

Research on the interactions between different erosion processes is limited. For instance, mass movement debris has significantly lower soil shear strength than the original loess deposits, thereby reducing soil erodibility. However, how this affects water erosion during subsequent rainstorms remains unknown. The dynamic changes in soil erodibility caused by mass movement have been rarely studied (Zhu et al., 2020). The role of mass movement in gully development needs further assessment. Additionally, the impact of runoff and sediment generated from intergully areas on erosion within the gully area is still unclear. There is a lack of detailed studies on the erosion, transport, and deposition of sediment in gully systems.

The Sediment Delivery Ratio (SDR) in the Loess Plateau region requires further research. Due to the steep terrain, SDR in the area has historically been assumed to be close to 1 (Gong & Xiong, 1979). However, with the implementation of numerous soil conservation measures, particularly check dams, sediment detention has become significant. As a result, the retention effects of these engineering works should be integrated into the SDR concept. Erosion models must therefore account for SDR at different spatial scales. Unfortunately, little to no research has been conducted on this topic to date.

Modeling erosion processes in the region remains a significant challenge (Li et al., 2017a, 2017b). Most process-based models primarily address hillslope erosion in intergully areas, while gully-area processes—such as tunnel erosion, mass movement, and gully erosion—are fundamentally different and largely underrepresented. Although a few models exist for these processes, they are typically empirical or semi-empirical. Given the prevalence of mass movement and gully erosion in the region, there is a pressing need to adapt and improve existing models to more accurately represent these complex dynamics.

References

Cai, Q. G. (1998). Research of rill initiation condition on loess hillslopes. *Journal of Sediment Research, 1*, 52–59. (in Chinese).

Cai, Q. G., Wu, S. A., Chen, H., Zhu, T. X., & Luk, S. H. (1990). Effects of soil crusts on runoff and soil loss. In Y. Z. Chen (Ed.), *Sino-Canadian symposium of Experiemntal research on the law of soil erosion in the loess plateau region, Western Shanxi*. China Water Conservancy and Hydropower Press. (In Chinese).

Chen, X. Y., Zhao, Y., Mi, H. X., & Mo, B. (2016). Estimating rill erosion process from eroded morphology in flume experiments by volume replacement method. *Catena, 136*, 135–140.

Chen, H., Zhang, X., Abla, M., Lü, D., Yan, R., Ren, Q., Ren, Z., Yang, Y., Zhao, W., Lin, P., & Liu, B. (2018). Effects of vegetation and rainfall types on surface runoff and soil erosion on steep slopes on the loess plateau, China. *Catena, 170*, 141–149.

Cheng, Y., Xia, J., Zhou, M., & Deng, S. (2024). Improved formula of sediment transport capacity and its application in the lower Yellow River. *Journal of Hydrology, 631*, 130812.

Dang, S., Liu, X., Yin, H., & Guo, X. (2020). Prediction of sediment yield in the middle reaches of the Yellow River basin under extreme precipitation. *Frontiers in Earth Science, 8*, 542686.

Derbyshire, E., Dijkstra, T. A., Smalley, I. J., & Li, Y. (1994). Failure mechanisms in loess and the effects of moisture content changes on remoulded strength. *Quaternary International, 24*, 5–15.

Eden, D. N., Qizhong, W., Hunt, J. L., & Whitton, J. S. (1994). Mineralogical and geochemical trends across the loess plateau, North China. *Catena, 21*(1), 73–90.

Fang, H. Y., Cai, Q. G., Chen, H., & Li, Q. Y. (2008). Effect of rainfall regime and slope on runoff in a gullied loess region on the loess plateau in China. *Environmental Management, 42*, 402–411.

Fang, H., Sun, L., & Tang, Z. (2015). Effects of rainfall and slope on runoff, soil erosion and rill development: An experimental study using two loess soils. *Hydrological Processes, 29*(11), 2649–2658.

Feng, X., Fu, B., Piao, S., Wang, S., Ciais, P., Zeng, Z., Lü, Y., Zeng, Y., Li, Y., Jiang, X., & Wu, B. (2016). Revegetation in China's loess plateau is approaching sustainable water resource limits. *Nature Climate Change, 6*(11), 1019–1022.

Food and Agriculture Organization (FAO). (1965). Soil erosion by water. In *Some measures for its control on cultivated lands*. Agricultural Paper 81.

Fu, B. (1989). Soil erosion and its control in the loess plateau of China. *Soil Use and Management, 5*(2), 76–82.

Fu, B., Liu, Y., Lü, Y., He, C., Zeng, Y., & Wu, B. (2011). Assessing the soil erosion control service of ecosystems change in the loess plateau of China. *Ecological Complexity, 8*(4), 284–293.

Fu, Y., Li, G., Zheng, T., Li, B., & Zhang, T. (2016). Impact of raindrop characteristics on the selective detachment and transport of aggregate fragments in the loess plateau of China. *Soil Science Society of America Journal, 80*(4), 1071–1077.

Gong, S. (1987). The role of reservoirs and silt-trap dams in reducing sediment delivery into the Yellow River. *Geografiska Annaler: Series A, Physical Geography, 69*(1), 173–179.

Gong, S. Y., & Xiong, G. S. (1979). The origin and the regional distribution of sediment of the Yellow River. *Yellow River, 1*, 7–11.

Guan, Y., Yang, S., Zhao, C., Lou, H., Chen, K., Zhang, C., & Wu, B. (2021). Monitoring long-term gully erosion and topographic thresholds in the marginal zone of the Chinese loess plateau. *Soil and Tillage Research, 205*, 104800.

Guo, Q., Cheng, C., Jiang, H., Liu, B., & Wang, Y. (2019a). Comparative rates of wind and water erosion on typical farmland at the northern end of the loess plateau, China. *Geoderma, 352*, 104–115.

Guo, W. Z., Luo, L., Wang, W. L., Liu, Z. Y., Chen, Z. X., Kang, H. L., & Yang, B. (2019b). Sensitivity of rainstorm-triggered shallow mass movements on gully slopes to topographical factors on the Chinese loess plateau. *Geomorphology, 337*, 69–78.

Guo, W. Z., Chen, Z. X., Wang, W. L., Gao, W. W., Guo, M. M., Kang, H. L., Li, P. F., Wang, W. X., & Zhao, M. (2020). Telling a different story: The promote role of vegetation in the initiation of shallow landslides during rainfall on the Chinese loess plateau. *Geomorphology, 350*, 106879.

Guo, Q., Wang, Z., Zhang, Q., Zhang, Q., Shen, N., Wu, B., Tian, N., & Liu, J. E. (2021). Quantifying sheet erosion rate on steep grassland in the loess region of China. *Archives of Agronomy and Soil Science, 67*(11), 1554–1565.

Han, F., Ren, L., Zhang, X., & Li, Z. (2016). The WEPP model application in a small watershed in the loess plateau. *PLoS One, 11*(3), e0148445.

Han, Y., Zheng, F. L., & Xu, X. M. (2017). Effects of rainfall regime and its character indices on soil loss at loessial hillslope with ephemeral gully. *Journal of Mountain Science, 14*, 527–538.

He, Z., Xiao, P., Yu, X., Hao, S., Jia, G., & Yang, C. (2022). A field study for the effects of grass cover, rainfall intensity and slope length on soil erosion in the loess plateau, China. *Water, 14*(14), 2142.

Hessel, R., & van Asch, T. (2003). Modelling gully erosion for a small catchment on the Chinese loess plateau. *Catena, 54*(1–2), 131–146.

Hessel, R., Jetten, V., Liu, B., Zhang, Y., & Stolte, J. (2003a). Calibration of the LISEM model for a small loess plateau catchment. *Catena, 54*(1–2), 235–254.

Hessel, R., Messing, I., Liding, C., Ritsema, C., & Stolte, J. (2003b). Soil erosion simulations of land use scenarios for a small loess plateau catchment. *Catena, 54*(1–2), 289–302.

Kariminejad, N., Rossi, M., Hosseinalizadeh, M., Pourghasemi, H. R., & Santosh, M. (2020). Gully head modelling in Iranian loess plateau under different scenarios. *Catena, 194*, 104769.

Kertész, Á., & Gergely, J. (2011). Gully erosion in Hungary, review and case study. *Procedia-Social and Behavioral Sciences, 19*, 693–701.

Laflen, J. M., Lane, L. J., & Foster, G. R. (1991). WEPP: A new generation of erosion prediction technology. *Journal of Soil and Water Conservation, 46*(1), 34–38.

Li, G., Luk, S., Cai, Q., & Liu, S. (1992). Study on infiltration law in WJG watershed, Lishi, China. In F. Wang (Ed.), *Soil erosion management and application of GIS in the loess plateau region, Western Shanxi* (p. 257). Science Press. (In Chinese with English abstract).

Li, Z., Zheng, F. L., Liu, W. Z., & Flanagan, D. C. (2010). Spatial distribution and temporal trends of extreme temperature and precipitation events on the loess plateau of China during 1961–2007. *Quaternary International, 226*(1–2), 92–100.

Li, P., Mu, X., Holden, J., Wu, Y., Irvine, B., Wang, F., Gao, P., Zhao, G., & Sun, W. (2017a). Comparison of soil erosion models used to study the Chinese loess plateau. *Earth-Science Reviews, 170*, 17–30.

Li, Z., Zhang, Y., Zhu, Q., Yang, S., Li, H., & Ma, H. (2017b). A gully erosion assessment model for the Chinese loess plateau based on changes in gully length and area. *Catena, 148*, 195–203.

Li, X. A., Wang, L., Yan, Y. L., Hong, B., & Li, L. C. (2019). Experimental study on the disintegration of loess in the loess plateau of China. *Bulletin of Engineering Geology and the Environment, 78*, 4907–4918.

Li, P., Chen, J., Zhao, G., Holden, J., Liu, B., Chan, F. K. S., Hu, J., Wu, P., & Mu, X. (2022a). Determining the drivers and rates of soil erosion on the loess plateau since 1901. *Science of the Total Environment, 823*, 153674.

Li, X. A., Sun, J., Ren, H., Lu, T., Ren, Y., & Pang, T. (2022b). The effect of particle size distribution and shape on the microscopic behaviour of loess via the DEM. *Environmental Earth Sciences, 81*(10), 290.

Liang, Z., Wu, Z., Yao, W., Noori, M., Yang, C., Xiao, P., Leng, Y., & Deng, L. (2019). Pisha sandstone: Causes, processes and erosion options for its control and prospects. *International Soil and Water Conservation Research, 7*(1), 1–8.

Liu, T. S. (1965). *Loess deposits in China*. Science Press.

Liu, B. Y., Nearing, M. A., & Risse, L. M. (1994). Slope gradient effects on soil loss for steep slopes. *Transactions of the ASAE, 37*(6), 1835–1840.

Liu, B. Y., Nearing, M. A., Shi, P. J., & Jia, Z. W. (2000). Slope length effects on soil loss for steep slopes. *Soil Science Society of America Journal, 64*(5), 1759–1763.

Liu, Y., Fu, B., Lü, Y., Wang, Z., & Gao, G. (2012). Hydrological responses and soil erosion potential of abandoned cropland in the loess plateau, China. *Geomorphology, 138*(1), 404–414.

Luk, S. H., & Cai, Q. G. (1990). Laboratory experiments on crust development and rainsplash erosion of loess soils, China. *Catena, 17*(3), 261–276.

Luk, S. H., Chen, H., Cai, Q. G., & Jia, Z. J. (1989). Spatial and temporal variations in the strength of loess soils, Lishi, China. *Geoderma, 45*(3–4), 303–317.

Luk, S. H., Cai, Q. G., Wang, G. P., Band, L., Fu, L. Y., & Zhu, T. X. (1992). Process-based erosion and sediment yield model in a small catchment in the hilly loess region. In F. T. Wang (Ed.), *Soil erosion management and application of GIS in the loess plateau region* (pp. 95–103). Science Press.

Luk, S. H., Cai, Q. G., & Wang, G. P. (1993). Effects of surface crusting and slope gradient on soil and water losses in the hilly loess region, North China. *Catena. Supplement (Giessen), 24*, 29–45.

Ma, B., Liu, G., Ma, F., Li, Z., & Wu, F. (2019). Effects of crop-slope interaction on slope runoff and erosion in the loess plateau. *Acta Agriculturae Scandinavica, Section B—Soil & Plant Science, 69*(1), 12–25.

Nachtergaele, J., & Poesen, J. (2002). Spatial and temporal variations in resistance of loess-derived soils to ephemeral gully erosion. *European Journal of Soil Science, 53*(3), 449–463.

Nachtergaele, J., Poesen, J., Steegen, A., Takken, I., Beuselinck, L., Vandekerckhove, L., & Govers, G. (2001). The value of a physically based model versus an empirical approach in the prediction of ephemeral gully erosion for loess-derived soils. *Geomorphology, 40*(3–4), 237–252.

Nearing, M. A., Foster, G. R., Lane, L. J., & Finkner, S. C. (1989). A process-based soil erosion model for USDA-water erosion prediction project technology. *Transactions of the ASAE, 32*(5), 1587–1593.

Qin, W., Guo, Q., Cao, W., Yin, Z., Yan, Q., Shan, Z., & Zheng, F. (2018). A new RUSLE slope length factor and its application to soil erosion assessment in a loess plateau watershed. *Soil and Tillage Research, 182*, 10–24.

Shen, H., Zheng, F., Wen, L., Lu, J., & Jiang, Y. (2015). An experimental study of rill erosion and morphology. *Geomorphology, 231*, 193–201.

Shen, H., Zheng, F., Wen, L., Han, Y., & Hu, W. (2016). Impacts of rainfall intensity and slope gradient on rill erosion processes at loessial hillslope. *Soil and Tillage Research, 155*, 429–436.

Shen, N., Wang, Z., Guo, Q., Zhang, Q., Wu, B., Liu, J., Ma, C., Delang, C. O., & Zhang, F. (2021). Soil detachment capacity by rill flow for five typical loess soils on the loess plateau of China. *Soil and Tillage Research, 213*, 105159.

Shi, Z. H., Fang, N. F., Wu, F. Z., et al. (2012a). Soil erosion processes and sediment sorting associated with transport mechanisms on steep slopes. *Journal of Hydrology, 454*, 123–130.
Shi, Z. H., Wu, F. Z., Yue, B. J., et al. (2012b). Effects of mulch cover rate on interrill erosion processes and the size selectivity of eroded sediment on steep slopes. *Soil Science Society of America Journal, 77*(1), 257–267.
Sun, L., Fang, H., Qi, D., Li, J., & Cai, Q. (2013). A review on rill erosion process and its influencing factors. *Chinese Geographical Science, 23*, 389–402.
Sun, W., Shao, Q., Liu, J., & Zhai, J. (2014). Assessing the effects of land use and topography on soil erosion on the loess plateau in China. *Catena, 121*, 151–163.
Sun, Q., Miao, C., Duan, Q., & Wang, Y. (2015). Temperature and precipitation changes over the loess plateau between 1961 and 2011, based on high-density gauge observations. *Global and Planetary Change, 132*, 1–10.
Sun, L., Zhou, J. L., Cai, Q., Liu, S., & Xiao, J. (2021). Comparing surface erosion processes in four soils from the loess plateau under extreme rainfall events. *International Soil and Water Conservation Research, 9*(4), 520–531.
Tang, K., Zhang, K., & Lei, A. (1998). Critical slope gradient for compulsory abandonment of farmland on the hilly loess plateau. *Chinese Science Bulletin, 43*, 409–412.
Thomas, A. W., & Welch, R. (1988). Measurement of ephemeral gully erosion. *Transactions of the ASAE, 31*(6), 1723–1728.
Tsunekawa, A., Liu, G., Yamanaka, N., & Du, S. (Eds.). (2014). *Restoration and development of the degraded loess plateau, China*. Springer.
Tuo, D., Xu, M., & Gao, G. (2018). Relative contributions of wind and water erosion to total soil loss and its effect on soil properties in sloping croplands of the Chinese loess plateau. *Science of the Total Environment, 633*, 1032–1040.
Valentin, C., Poesen, J., & Li, Y. (2005). Gully erosion: Impacts, factors and control. *Catena, 63*(2–3), 132–153.
Wang, G. (1998). Summary of rill erosion study. *Soil and Water Conservation in China, 8*, 23–26. (in Chinese).
Wang, G., Wu, B., & Li, T. (2007). Digital yellow river model. *Journal of Hydro-Environment Research, 1*(1), 1–11.
Wang, Q., Bai, Y., Gao, H., He, J., Chen, H., Chesney, R. C., Kuhn, N. J., & Li, H. (2008a). Soil chemical properties and microbial biomass after 16 years of no-tillage farming on the loess plateau, China. *Geoderma, 144*(3–4), 502–508.
Wang, X., Gao, H., Tullberg, J. N., Li, H., Kuhn, N., McHugh, A. D., & Li, Y. (2008b). Traffic and tillage effects on runoff and soil loss on the loess plateau of northern China. *Soil Research, 46*(8), 667–675.
Wang, H. B., Zhou, B., Wu, S. R., Shi, J. S., & Li, B. (2011). Characteristic analysis of large-scale loess landslides: A case study in Baoji City of loess plateau of Northwest China. *Natural Hazards and Earth System Sciences, 11*(7), 1829–1837.
Wang, Z. J., Jiao, J. Y., Rayburg, S., Wang, Q. L., & Su, Y. (2016). Soil erosion resistance of "grain for green" vegetation types under extreme rainfall conditions on the loess plateau, China. *Catena, 141*, 109–116.
Wang, D., Wang, Z., Zhang, Q., Zhang, Q., Tian, N., & Liu, J. E. (2018). Sheet erosion rates and erosion control on steep rangelands in loess regions. *Earth Surface Processes and Landforms, 43*(14), 2926–2934.
Wang, L., Li, X. A., Li, L. C., Hong, B., Yao, W., Lei, H. N., & Zhang, C. (2020). Characterization of the collapsible mechanisms of Malan loess on the Chinese loess plateau and their effects on eroded loess landforms. *Human and Ecological Risk Assessment: An International Journal, 26*(9), 2541–2566.
Wang, J., Zhang, Y., Deng, J., Yu, S., & Zhao, Y. (2021). Long-term gully erosion and its response to human intervention in the tableland region of the chinese loess plateau. *Remote Sensing, 13*(24), 5053.

Wang, L., Li, X. A., Qin, B., Zheng, H., & Zheng, Z. (2024). Genesis and evolution mechanism of loess tunnels in the loess plateau, China. *Catena, 239*, 107919.

Wei, W., Chen, L. D., Fu, B. J., et al. (2007). The effect of land uses and rainfall regimes on runoff and soil erosion in the semi-arid loess hilly area, China. *Journal of Hydrology, 335*(3/4), 247–258.

Wei, W., Jia, F., Yang, L., Chen, L., Zhang, H., & Yu, Y. (2014). Effects of surficial condition and rainfall intensity on runoff in a loess hilly area, China. *Journal of Hydrology, 513*, 115–126.

Wu, Y., & Cheng, H. (2005). Monitoring of gully erosion on the loess plateau of China using a global positioning system. *Catena, 63*(2–3), 154–166.

Wu, B., Wang, Z., Zhang, Q., Shen, N., & Liu, J. (2017). Modelling sheet erosion on steep slopes in the loess region of China. *Journal of Hydrology, 553*, 549–558.

Wu, B., Wang, Z., Zhang, Q., Shen, N., & Liu, J. (2019a). Evaluating and modelling splash detachment capacity based on laboratory experiments. *Catena, 176*, 189–196.

Wu, B., Wang, Z. L., Zhang, Q. W., Shen, N., & Liu, J. (2019b). Response of soil detachment rate by raindrop-affected sediment-laden sheet flow to sediment load and hydraulic parameters within a detachment-limited sheet erosion system on steep slopes on loess plateau, China. *Soil and Tillage Research, 185*, 9–16.

Xin, Z., Yu, X., Li, Q., & Lu, X. X. (2011). Spatiotemporal variation in rainfall erosivity on the Chinese loess plateau during the period 1956–2008. *Regional Environmental Change, 11*, 149–159.

Xu, X. Z., Liu, Z. Y., Wang, W. L., Zhang, H. W., Yan, Q., Zhao, C., & Guo, W. Z. (2015a). Which is more hazardous: Avalanche, landslide, or mudslide? *Natural Hazards, 76*, 1939–1945.

Xu, X. Z., Liu, Z. Y., Xiao, P. Q., Guo, W. Z., Zhang, H. W., Zhao, C., & Yan, Q. (2015b). Gravity erosion on the steep loess slope: Behavior, trigger and sensitivity. *Catena, 135*, 231–239.

Xu, X. Z., Zhang, H. W., Wang, W. L., Zhao, C., & Yan, Q. (2015c). Quantitative monitoring of gravity erosion using a novel 3D surface measuring technique: Validation and case study. *Natural Hazards, 75*, 1927–1939.

Xu, X. Z., Guo, W. Z., Liu, Y. K., Ma, J. Z., Wang, W. L., Zhang, H. W., & Gao, H. (2017). Landslides on the loess plateau of China: A latest statistics together with a close look. *Natural Hazards, 86*, 1393–1403.

Xu, Q., Li, M., Jiang, X., Zhang, Z., Jiao, J., Jian, J., Li, J., Yan, X., Liang, Y., Chen, T., & Chen, Y. (2022). Response of rill erosion to rainfall types and maintenance on the loess plateau: Implications for road erosion control. *Catena, 219*, 106642.

Yang, T., Xu, C. Y., Zhang, Q., Yu, Z., Baron, A., Wang, X., & Singh, V. P. (2012). DEM-based numerical modelling of runoff and soil erosion processes in the hilly–gully loess regions. *Stochastic Environmental Research and Risk Assessment, 26*, 581–597.

Yu, Y., Zhao, W., Martinez-Murillo, J. F., & Pereira, P. (2020). Loess plateau: From degradation to restoration. *Science of the Total Environment, 738*, 140206.

Zhang, H. W., & Zhang, Q. (1992). Formula for the sediment transport capacity of the Yellow River (in Chinese). *Yellow River, 11*, 7–9.

Zhang, K., Li, S., Peng, W., & Yu, B. (2004). Erodibility of agricultural soils on the loess plateau of China. *Soil and Tillage Research, 76*(2), 157–165.

Zhang, J., Yang, M., Sun, X., & Zhang, F. (2018). Estimation of wind and water erosion based on slope aspects in the crisscross region of the Chinese loess plateau. *Journal of Soils and Sediments, 18*, 1620–1631.

Zhang, J., Yang, M., Deng, X., Liu, Z., & Zhang, F. (2019). The effects of tillage on sheet erosion on sloping fields in the windwater erosion crisscross region of the Chinese loess plateau. *Soil and Tillage Research, 187*, 235–245.

Zhao, G., Mu, X., Wen, Z., Wang, F., & Gao, P. (2013). Soil erosion, conservation, and eco-environment changes in the loess plateau of China. *Land Degradation & Development, 24*(5), 499–510.

Zheng, F. (1989). Critical slope length and slope gradient for the occurrence of rill erosion. *Soil and Water Conservation in China, 8*, 23–24. (in Chinese).

Zheng, F. L. (2006). Effect of vegetation changes on soil erosion on the loess plateau. *Pedosphere, 16*(4), 420–427.

Zheng, F. L., & Tang, K. L. (1997). Rill erosion process on steep slope land of the loess plateau. *International Journal of Sediment Research, 12*(1), 52–59.

Zheng, H., Miao, C., Wu, J., Lei, X., Liao, W., & Li, H. (2019). Temporal and spatial variations in water discharge and sediment load on the loess plateau, China: A high-density study. *Science of the Total Environment, 666*, 875–886.

Zheng, F., Zhang, X. J., Wang, J., & Flanagan, D. C. (2020). Assessing applicability of the WEPP hillslope model to steep landscapes in the northern loess plateau of China. *Soil and Tillage Research, 197*, 104492.

Zhou, P. H., & Wang, Z. L. (1987). Soil erosion storm rainfall standard in the loess plateau. *Bulletin of Soil and Water Conservation, 1*, 38–44. (in Chinese).

Zhu, T. X. (1989). Geomorphic thresholds of mass wasting in the loess area. In Y. Z. Chen (Ed.), *Coarse silt source and erosion mechanisms in the Yellow River* (pp. 77–109). Meteorology Press. (in Chinese).

Zhu, T. X. (1997). Deep-seated., complex tunnel systems—A hydrological study in a semi-arid catchment, loess plateau, China. *Geomorphology, 20*(3–4), 255–267.

Zhu, T. X. (2003). Tunnel development over a 12 year period in a semi-arid catchment of the loess plateau, China. *Earth Surface Processes and Landforms: The Journal of the British Geomorphological Research Group, 28*(5), 507–525.

Zhu, T. X. (2012). Gully and tunnel erosion in the hilly loess plateau region, China. *Geomorphology, 153*, 144–155.

Zhu, T. X. (2016). Effectiveness of conservation measures in reducing runoff and soil loss under different magnitude–frequency storms at plot and catchment scales in the semi-arid agricultural landscape. *Environmental Management, 57*, 671–682.

Zhu, T. X., & Xu, X. Z. (2021). *Watershed erosion processes*. Springer.

Zhu, T. X., & Zhu, A. X. (2014). Assessment of soil erosion and conservation on agricultural sloping lands using plot data in the semi-arid hilly loess region of China. *Journal of Hydrology: Regional Studies, 2*, 69–83.

Zhu, T., Cai, Q. G., & Zeng, B. Q. (1997). Runoff generation on a semi-arid agricultural catchment: Field and experimental studies. *Journal of Hydrology, 196*(1–4), 99–118.

Zhu, T. X., Luk, S. H., & Cai, Q. G. (2002). Tunnel erosion and sediment production in the hilly loess region, North China. *Journal of Hydrology, 257*(1–4), 78–90.

Zhu, T., Xu, X., & Zhu, A. (2019). Spatial variation in the frequency and magnitude of mass movement in a semiarid, complex-terrain agricultural watershed on the L oess P lateau of C hina. *Land Degradation & Development, 30*(9), 1095–1106.

Zhu, T., Xu, X., Zhu, A. X., & Wang, X. (2020). Assessment of soil shear strength in gully systems: A case-study in the Wangjiagou (WJG) watershed on the loess plateau of China. *Land Degradation & Development, 31*(17), 2648–2661.

Zhu, T., Xu, X., Zhu, T., & Xu, X. (2021). *Sheet and rill erosion* (pp. 21–39). Springer.

Zhu, R., Yu, Y., Zhao, J., Liu, D., Cai, S., Feng, J., & Rodrigo-Comino, J. (2023). Evaluating the applicability of the water erosion prediction project (WEPP) model to runoff and soil loss of sandstone reliefs in the loess plateau, China. *International Soil and Water Conservation Research, 11*(2), 240–250.

Zhu, T. X., Luk, S. H., Cai, Q. G., Gao, P., Wei, Z. P., & Tailor, H. (1992). Runoff and sediment processes in the YDG catchment. In F. T. Wang (Ed.), *Soil Erosion management and application of GIS in the Loess Plateau Region, Western Shanxi* (pp. 95–103). Science Press.

Chapter 5
Soil Conservation and Sustainability

Contents

5.1 Introduction

The Loess Plateau contributes over 80% of the sediment in the Yellow River, making it the key area for soil erosion control in the Yellow River basin. Reducing soil, water, and nutrient loss from agricultural lands is crucial for sustaining agricultural productivity, decreasing sediment buildup and downstream flood risks, and preventing the deterioration of water quality in rivers and other water bodies. It is a key strategy for promoting sustainable economic development, reducing poverty and improving the standard of living for local communities, and reversing ecosystem degradation in the region. Three main types of soil conservation measures have been implemented in the region: engineering, biological, and agronomic practices since the 1950s. By 2018, conservation efforts covered 48% of the areas most affected by severe erosion within the Yellow River basin. As a result, the average annual sediment discharge into the Yellow River from 2001 to 2018 decreased by 85% compared to the levels recorded between 1919 and 1959. There have been numerous studies conducted in the region to evaluate the effectiveness of different conservation measures in reducing soil and water loss (Chen et al., 2007; Zhao et al., 2013). A variety of methods have been adopted, including field monitoring plots under

T. Zhu, *Watershed Environmental Degradation and Sustainable Development*,
https://doi.org/10.1007/978-3-032-02448-0_5

natural and rainfall simulation experiments (Cai, 2001; Sun et al., 2014; Zhu & Zhu, 2014), remote sensing techniques (Yang et al., 2020), comparing watersheds with and without conservation measures (Zhu, 2016), and modeling approaches (Fu et al., 2011; Li et al., 2017). In this chapter, we will begin by exploring the types of soil conservation measures implemented in the Loess Plateau, and then discuss the effectiveness of comprehensive conservation efforts undertaken at the scale of watersheds.

5.2 Soil Conservation Measures

5.2.1 Engineering Measures

Terrace land is one of the most important engineering soil conservation measures in the Loess Plateau region, with a total area of 9.02 ± 0.32 million hectares, or about 14% of the total area (Jiao et al., 1999; Lu et al., 2023). Converting slope land to terrace land can greatly enhance rainfall infiltration and soil moisture, reducing surface runoff and soil erosion while also boosting crop yields and promoting sustainable agriculture. Multi-year field observations in the Wangjiagou experimental watershed in Shanxi Province indicate that terrace land reduces average annual runoff by 59.9% and average annual soil loss by 92.4% compared to conventional cropland (Zhu & Zhu, 2014) (Fig. 5.1). Field observations showed that soil moisture content at 0–1 m soil depth range on terrace land can be 4.39% to 4.89% higher than slope farmland (Xu et al., 2021). A three-year continuous monitoring in the western Loess Plateau indicated that the soil organic content of the terrace land increased by 26%, total N by 8%, total P by 4%, fast acting N by 12%, and fast acting P by 20% (Liu et al., 2011). A long-term study by Jiao et al. (1999) demonstrated that the total nitrogen and organic content increased by 150 and 78%, respectively over a 20-year period (1961–1981). The retention of water and nutrients leads to the improvement of land productivity. The crop production from the terrace land (older than 3 years) was 27% higher than slope farmland (Liu et al., 2011). Such a difference in crop yield between terrace land and slope farmland is even greater in dry years. In addition, retaining rainwater by upslope terraces significantly reduces erosion downslope and in gullies.

Although terraced lands offer these benefits, they also come with certain drawbacks. Constructing terraces involves an initial expense, and the significant soil disturbance during this process can lead to increased soil loss in the first year post-construction. Converting slope land to terrace land is generally confined to the areas above the gully-edge line, as it's not cost-effective on the steeper slopes below. Over time (20+ years), soil can become compacted from continuous cultivation, potentially forming plough pans beneath the terrace surface that limit water infiltration and increase surface runoff and soil loss during heavy rainstorms. Soil piping and tunnel formation along terrace surfaces and edges may also occur, which can

Fig. 5.1 Terrace land in the Wangjiagou watershed, Shanxi Province

lead to terrace damage (Zhu, 2003). Additionally, terraces'effectiveness in conserving soil and water decreases under extreme storm conditions, with older terraces particularly vulnerable to damage during very large storms.

Another widely implemented engineering measure in the Loess Plateau is the construction of check dams along valley bottoms (Fig. 5.2). By 2005, approximately 122,000 check dams have already been built in this region (Xu et al., 2013; Wang et al., 2014), with a plan to construct 163,000 more (Jin et al., 2012). According to Fang et al. (2023), the check dams in the Loess Plateau have intercepted approximately 10.2 ± 0.6 Pg of eroded sediment between 1970 and 2020, accounting for 46% of the Yellow River's sediment load. Check-dam lands are nutrient-rich and maintain high soil moisture, often transforming into high-yield croplands. Nearly all the sediment-borne nutrients lost from the upper slope land above the check-dam can be captured by the dam. Hamilton and Luk (1992) estimated that 98% of nutrient nitrogen loss was associated with sediment. They observed that the total N loss was 76.4 kg/ha in 1987 and 44. Kg/ha in 1988 in the Yangdaogou catchment, a subwatershed of the Wangjiagou watershed. Check-dam lands also serve as carbon storage and sequential structure in the region. Fang et al. (2023) estimated that 21.6 ± 9.9 Tg of organic carbon was buried over the past 50 years by check dams in the region. Check dams retain runoff generated from the catchment and thereby increase soil moisture on the check-dam land. Observations from the Wangjiagou watershed in 2006 revealed an average soil moisture of 16.95% on check-dam land,

Fig. 5.2 Cropland created behind a check-dam in the Wangjiagou watershed

compared to 9.49% on terrace land and 8.67% on slope cropland. This means that check-dam land has 1.786 times the soil moisture of terrace land and 1.955 times that of slope cropland. The differences are even more pronounced in dry years, where moisture becomes the primary factor limiting crop yields. The increased moisture and nutrient content of the check-dam land result in notably higher crop yields—2-3 times that of terraced cropland and 6–10 times that of sloped cropland (Wang et al., 2011). Overall, check-dam construction has created 3200 km^2 of high-yield farmland on the Loess Plateau (Xu et al., 2004).

Check dams also can reduce mass movement and gully development (Xu et al., 2020). As sedimentation extends upstream and upslope from the end of the check-dam, the topographic relief above the dam decreases, helping to protect the channel's sideslopes from mass movement and preventing gully incision and further gully development.

The primary concern with check dams is their safety, as failures often occur during extreme storm events. After decades of operation, 3025 large and 2257 medium-sized check dams in the Loess Plateau region are considered dangerous and pose safety risks, seriously threatening downstream properties and lives. This creates an urgent need for reinforcing these hazardous check dams (Guo et al., 2021).

5.2.2 *Biological Measures*

The major types of biological measures implemented in the Loess Plateau include grass, shrub, tree, and biological crust. A large-scale revegetation effort has been initiated in the Loess Plateau, particularly since the launch of the "Grain for Green Program" in 1999. Slope lands with gradient greater than 25° are required to convert agricultural land back into forest or woodland to promote environmental restoration and sustainability. About 16,000 km^2 of farmland has been converted to vegetation land, leading to an increase of vegetation coverage by 25% between 1999 and 2013 (Feng et al., 2016; Zhang et al., 2022). It is well known that vegetation can reduce soil loss, increase carbon storage, and restore degraded ecosystems.

Grassland occupies more than 1/3 of the Loess Plateau area. In the Loess Plateau, the natural grassland is dominated by *Stipa bungeana* Trin. and *Artemisia sacrorum Ledeb, Stipa breviflora, and Thymis mongolicus whereas the artificial gransland is dominated by Medicago sativa L., Astragalus adsurgens Pall., Panicum virgatum* L., Artemisia scoparia, *among others* (Chen et al., 2010; Zhang et al., 2016;Wang et al., 2013). Grasses help prevent soil erosion by anchoring soil particles with their roots. Liu et al. (2022) indicated that shear strength was linearly correlated with root density of grasses and that root tensile strength decreased with root diameter with a power law relationship. Grass enhances soil structure and infiltration capacity, helping to reduce surface runoff. It also slows down water flow on the surface, which in turn reduces surface erosion. Wang et al. (2012) reported that grasslands exhibit a higher rate of infiltrability and deeper wetting front than croplands, absorbing 60% more water despite lower soil moisture content, suggesting higher evaporation from grasslands.

Several factors influence the efficiency of soil erosion control on grasslands, with vegetation coverage being a key determinant of their effectiveness in soil and water conservation. Research indicates that grass coverage needs to exceed 60–70% of the ground area to effectively control soil erosion (Wang et al., 1992; Li et al., 2005). Liu et al. (2020) found that when grass coverage falls below 40%, grasslands can produce more runoff than farmlands, as the deeper tilling in farmlands improves infiltration. With regard to slope gradient, the optimal reduction efficiency of grasslands occurs between 15–20 degrees, with effectiveness gradually declining beyond this range. Most studies indicate that while the runoff reduction efficiency of grasslands decreases during extremely heavy storms (e.g., Guo et al., 2019; Wang et al., 2012), their ability to reduce soil loss remains strong. For instance, Bai (1997) observed that grasslands can still reduce soil loss by 70–90% during extreme rainstorms, even with daily precipitation of 120 mm or events with an 80-year recurrence interval. Data from erosion plots in the Wangjiagou watershed showed that grasslands produced a total of 22.48 mm of runoff and 3912 t/km^2 of soil loss, compared to 146.9 mm of runoff and 57,335 t/km^2 of soil loss over an eight-year period, achieving reduction rates of 85% for runoff and 93.2% for soil loss. In most individual storm events, grasslands can reduce soil loss by more than 90%. However, during one extreme event with 104 mm of rainfall, the reduction rate dropped to

67.4%. Fu et al. (2010) reported that, compared to slope croplands, grasslands can reduce average runoff and sediment yield by 18% and 76%, respectively. In contrast, Bai et al. (1993) found through field experiments that damage to natural grasslands could cut infiltration rates in half, leading to a significant increase in surface runoff by 1273 to 3050 m^3/km^2 and sediment yield by 500 to 1700 t/km^2.

Shrubs are also crucial in mitigating soil erosion in the Loess Plateau region, especially in gully areas (Fig. 5.3). Key shrub species in this area include *Caragana korshinskii*, *Spiraea pubescens*, and *Hippophae rhamnoides*. These shrubs contribute to improving soil quality over time by increasing soil organic content and enhancing soil aggregates. For example, *Caragana korshinskii* (Fig. 5.4), a common shrub in the region, shows a rise in soil organic carbon (SOC) content as it matures. Zeng et al. (2017) reported SOC concentrations (g/kg) in the 0–20 cm soil layer for 10, 20, and 30 years as 10.79, 12.66, and 13.85, respectively. Research indicates that *Caragana korshinskii* may build larger nutrient reserves than natural grasslands (Xu et al., 2014). Shrubs can significantly reduce runoff (by 57.5–71.2%) and soil loss (by 86–95.5%) compared to bare land on the Loess Plateau (Cai, 2001), through monitoring field plots under natural conditions (Fig. 5.5).

While both grass and shrub play important roles in controlling soil erosion, their relative effectiveness is a subject of debate. Wei et al. (2007) advocate for grasslands as the most effective solution for erosion control, with shrublands and woodlands

Fig. 5.3 Shrubs in the gully area of the Wangjiagou watershed

Fig. 5.4 *Caragana korshinskii planted* in the Wangjiagou watershed

Fig. 5.5 Shrub plots in the experimental station of Shanxi Institute of Soil and Water Conservation

serving as useful complements. Conversely, Zhu et al. (2021) highlight that shrub communities are more effective in reducing erosion, while grass communities—especially those with deep taproot systems—are better suited for managing runoff on steep slopes. Chen et al. (2018) stress that dense ground cover in both grassland and shrubland can significantly minimize runoff and soil loss. However, shrubs, with their thicker and deeper roots, are particularly effective at stabilizing slopes and reducing the risk of mass movements.

Water availability is a major challenge for maintaining biological communities in semi-arid regions. Shrubs, with their deep root systems, access soil moisture from deeper layers that have accumulated over previous years, which can result in the formation of a deep dry soil layer. This contrasts with grasses and crops, which primarily depend on shallow soil moisture that is replenished annually (Wang et al., 2009). Studies have shown that various shrub species significantly influence soil water storage at depths of 100–500 cm below the surface (Liu et al., 2016). Additionally, tunnel systems can form in shrublands (Zhu, 2003).

Significant efforts have been made to promote tree planting in the Loess Plateau, particularly with the introduction of the "Grain for Green" program by the Chinese government in 1999. The region hosts a diverse range of tree species, including Chinese pine, Mongolian pine, black locust, Cathay poplar, and Siberian elm (Wang et al., 2019). Techniques like fish-scale pits, contour trenching, and the use of biodegradable plastics have been adopted for planting trees (Fu et al., 2010). These methods aim to improve water retention by modifying the landscape's small-scale topography or by reducing the evaporation of soil moisture. For example, studies have shown a notable 18% increase in soil moisture within fish-scale pits compared to uncovered slopes (Wang et al., 2015). The application of biodegradable plastic around plantings has been shown to decrease evaporation and increase soil moisture, thereby improving the survival rates of trees (Cao et al., 2008). The development of secondary forests has led to a marked increase in soil organic matter and total nitrogen, particularly in the top 60 cm of soil, while the bulk density of the soil has shown a decrease, especially in the upper 20 cm layer (Deng et al., 2013). Numerous studies in the area have highlighted the effectiveness of forests in curbing soil erosion. For instance, Zheng (2006) reported a significant drop in soil erosion in the Ziwulin section of the Loess Plateau following the introduction of secondary forests. The canopy and leaf litter of trees play a crucial role in intercepting rainfall and reducing the impact energy of raindrops, with the litter layer also serving to shield the soil from erosion and slow the flow of water. Experimental data reveals that the presence of a litter layer alone can decrease runoff by 18.6%–25.9% and soil erosion by 58.5%–78.3% compared to bare soil (Xia et al., 2019). Liu et al. (2020) reported a reduction in soil loss of over 90% when compared to grassland areas. Despite the benefits of reforestation, the long-term viability of a large-scale reforestation efforts remains a topic of discussion and concern. Reforestation can result in increased evapotranspiration (ET). Model simulations estimate that the Grain for Green program has increased evaporative water consumption across the Loess Plateau by about 3.1 billion m^3 per year (or 4.90 mm/year) (Shao et al., 2019). Forests generally need a minimum annual amount of about 400–500 mm in mid-latitude climates to thrive. Thus, it is recommended that tree-planting may be more appropriate in areas with high amount of annual precipitation in the Loess Plateau region.

Biological crusts, or biocrusts, represent an underexplored biological approach to mitigating soil erosion in the Loess Plateau region (Fig. 5.6). These crusts are composed of a community of organisms including cyanobacteria, algae, fungi, bacteria, lichens, and mosses, which either reside on the soil's surface or within a few centimeters of the top layer (Belnap et al., 2016). They contribute to an increase in soil organic matter, enhance water infiltration rates, improve soil structure, and protection against soil erosion. Research by Gao et al. (2017) demonstrated that areas dominated by moss biocrusts had soil organic matter content four times higher than that of bare soil, while cyanobacteria-dominated biocrusts exhibited 1.5 times increase. Furthermore, biocrusts have been shown to significantly decrease soil erosion by up to 90% through modifications in soil properties.

Studies have been conducted to compare the effectiveness of different vegetation types in reducing soil loss and their impact on soil properties. A six-year study on soil erosion plots in the Wangjiagou watershed compared forest, grasses, alfalfa, and slope cropland. The mean annual soil loss was 667.7 g/m^2 for forest, 489.2 g/m^2 for grasses, and 1575 g/m^2 for alfalfa, representing 10.6%, 6.9%, and 25.4% of the soil loss from slope cropland, which experienced 6279 g/m^2 of soil loss. However, during the first year of planting, soil loss from the vegetated plots exceeded that of the slope cropland, with 3690 g/m^2 from forest, 3903.9 g/m^2 from grassland, and 2900 g/m^2 from alfalfa, compared to 2517.6 g/m^2 from the slope cropland plot. This increase was attributed to soil disturbance during planting and low vegetation cover

Fig. 5.6 Biological crust growing on the debris of loess fall

during the early growth stage (Zhu & Zhu, 2014). A long term observation of soil moisture content (SMC) over a period of 12-year on three types of revegationed lands and cropland shows that the mean soil moisture content at depth of 0–4 m was 17.07%, 15.74%, 11.99% and 11.66% on cropland, fallow land, grassland and shrubland, respectively (Zhao et al., 2017). They also found that among the three revegetated lands (fallow land, grassland, and shrubland), fallow land had stronger stability and higher value of SWC and was thus recommended as the best revegetation type. Zhu et al. (2021) investigated soil quality recovery across different vegetation types, including two shrublands (*Caragana korshinskii* and *Hippophae rhamnoides*), two grasslands (*Astragalus adsurgens* and *Panicum virgatum*), and two species from croplands left to recover naturally (*Artemisia capillaris* and *Heteropappus altaicus*). The study found that natural recovery of abandoned croplands had the highest capacity for improving soil quality—measured by soil organic carbon, total nitrogen, total phosphorus, available nitrogen, available phosphorus, moisture content, and microbial biomass—among all the revegetation types studied in the Loess Plateau.

Given that annual precipitation on the Loess Plateau varies from 200 mm in the northwest to 750 mm in the southeast, implementing revegetation plans in the Loess Plateau should carefully consider the spatial variability of natural conditions, particularly precipitation. Selecting vegetation types should focus on ensuring sustainable water use while maximizing economic and social benefits.

5.2.3 Agronomic Measures

Effective agronomic practices can enhance soil properties, reduce soil erodibility, and decrease surface runoff and soil loss. Various practices have been implemented in the region, such as no-till or reduced-till farming, breaking up soil crusts and plow pans, and covering cropland with crop residues, among others.

A 16-year study by Wang et al. (2008) demonstrated that no-tillage with straw cover increased soil organic carbon (SOC) by 21.7% and total nitrogen (TN) by 51.0% at a depth of 0–10 cm, and available phosphorus (P) by 97.3% at 0–5 cm, compared to traditional tillage. Additionally, soil microbial biomass carbon (C) and nitrogen (N) increased by 135.3% and 104.4%, respectively, in the 0–10 cm layer. No-tillage also reduced bulk density in the 20–30 cm layer and increased macroporosity and saturated hydraulic conductivity at the 15–30 cm soil depth, although no significant differences were observed in the 0–15 cm surface layers between no-tillage with straw cover and traditional tillage (He et al., 2009). Moreover, straw mulching plays a more crucial role than no-tillage in enhancing crop yield in the region (Xiao et al., 2019).

Local farmers often use hoes to break soil crusts. Loess soils are particularly susceptible to crust formation due to their high silt content, low organic matter, and poor aggregate stability. When raindrops impact these soils, soil aggregates break down easily, causing fine particles to settle on the surface and form a dense, thin crust that greatly reduces infiltration capacity. The formation of crusts is mainly a physical and mechanical process rather than a chemical one in this region. This soil crust can increase surface runoff by decreasing the infiltration rate. To evaluate the impact of this practice on runoff generation and soil loss, we monitored soil and water loss on two set of contrasting plots over a two-year period. In one set of replicate plots, we broke the soil crust after each storm, while the other set was left undisturbed. The results indicated minimal differences in soil moisture content between the crusted and uncrusted plots at depths of 0–10 cm, 10–20 cm, and 20–30 cm. Water loss from uncrusted plots was slightly lower than that from crusted plots. However, soil loss from the broken crusts was significantly higher than from the intact crusted plots. The small difference in soil moisture and runoff between the broken and intact crust plots can be attributed to the rapid reformation of crusts on the disturbed surfaces. Yet, the disruption of soil surfaces in the broken crust plots led to severe soil loss, particularly during heavy rainstorms. Our study raises questions about the effectiveness of breaking crusts on farmland as a means to reduce soil loss (Zhu, 2002).

Plow pans form after prolonged cultivation, especially on terraced or gently sloped farmland. When the slope gradient is greater than 15 degrees, the plow pan becomes less pronounced due to the downslope movement of displaced soil during plowing. Repeated plowing compacts the soil, significantly reducing the porosity of the plow pan layer compared to the layers above and below. Typically found 15–20 cm below the surface, the plow pan has a bulk density of 1.26 to 1.43 g/cm^3, while the layers above (0–15 cm) range from 1.13 to 1.26 g/cm^3, and the layers below (25–30 cm) range from 1.19 to 1.31 g/cm^3 (Cai et al., 1990). The plow pan's reduced water transmission rate can cause water to accumulate above it, increasing the risk of surface runoff. Studies by Cai et al. (1990) using rainfall simulation experiments found that plots with intact plow pans produced 1.5 to 1.8 times more runoff and soil loss compared to plots with broken plow pans. Increasing the depth of plowing can break the plow pan layer, improving water transmission and soil moisture in the subsoil, thereby reducing surface runoff and soil erosion (Zhu et al., 1997). Breaking up the plow pan can also enhance soil moisture and fertility, leading to higher crop yields.

The Loess Plateau region has also adopted practices like crop rotation, intercropping, and multiple cropping. Field experiments conducted by Li et al. (2002) over three years in the southern Loess Plateau revealed that crop rotation significantly enhances soil water-stable aggregates and improves the efficient use of rainfall compared to a monoculture of winter wheat (*Triticum aestivum* L.). Huang et al. (2003) recommended that a 3-year rotation of corn, wheat, and wheat–millet was found to offer the best potential for improving cropping practice in the region.

5.3 Comprehensive Management of Soil and Water Loss at the Small watershed Scale

A small watershed acts as a fundamental unit for applying comprehensive strategies to control soil, water, and nutrient loss. Changing inappropriate landuse within a watershed is essential for controlling soil erosion. The spatial arrangement of engineering, biological, and agricultural practices within the watershed should focus on sustainable development, aiming to optimize the environmental, economic, and social benefits.

Watersheds in the Loess Plateau consist of inter-gully and gully areas, divided by the gully edge line. The inter-gully areas, located above the gully edges, have slopes ranging from gentle to moderately steep. Converting these natural slopes into terraces is a cost-effective method to enhance infiltration, reduce surface runoff, and improve agricultural yields, especially when compared to terracing the steep slopes of gullies. By minimizing runoff from inter-gully areas, erosion on the gully slopes can be significantly reduced. Deep plowing on terrace lands can enhance water infiltration and disrupt the plow pan, thereby improving soil moisture and crop yields. The practice of returning crop residue to the terrace lands can also decrease evaporation and boost soil fertility. Given the steepness of gully slopes, biological approaches such as planting trees, shrubs, and grasses are more suitable. The selection of plant types and species should align with the local climatic conditions. Generally, in areas where the average yearly rainfall is below 400 mm, afferestation is not advisable due to insufficient rainfall to support forest growth sustainably. In such cases, species of grasses and shrubs that are resistant to drought conditions should be considered instead. In regions receiving more rainfall annually, a variety of trees, shrubs, and grasses can be successfully cultivated, with an emphasis on selecting economically beneficial plants. Additionally, restricting grazing in the gully areas allows vegetation to regenerate naturally over time (Fig. 5.7). At the gully bottom, constructing a series of check dams further aids in capturing eroded sediments from the hillslopes. The flat areas developed behind these dams gradually transform into fertile agricultural land with higher soil nutrient and moisture levels than slope farmlands and terraces.

To evaluate the impact of comprehensive soil conservation strategies at the small watershed level, the Shanxi Institute of Soil and Water Conservation (SISWC) conducted a 15-year field monitoring program (1956–1980) in two neighboring catchments, YDG and CCZG, with drainage areas of 0.206 km^2 and 0.193 km^2, respectively. Both catchments have similar topographies, featuring steep slopes, with slopes steeper than 30° covering 48.1% of YDG and 41.0% of CCZG (Zhu, 2016). Before 1956, both catchments had similar land use, with slope croplands dominating the inter-gully areas and the gully areas being either barren or sparsely covered with shrubs. Post-1956, the SISWC implemented various conservation measures in CCZG, including converting upper slope croplands into terraces and planting trees and shrubs on lower slopes. Grazing was also prohibited to promote natural grass regeneration on the gully slopes. These conservation efforts

Fig. 5.7 Implementing grazing restrictions in gully areas and encouraging vegetation growth on the Loess Plateau

encompassed 78.3% of the CCZG area. Meanwhile, YDG's land use remained unchanged, allowing for a direct comparison between the managed CCZG and the unmanaged YDG, enabling them to evaluate the effectiveness of soil conservation measures on soil erosion reduction. The results showed that CCZG experienced a 59.1% reduction in average annual soil loss and a 55.9% reduction in runoff compared to YDG over the 15-year period. Similar trends were observed across experimental watersheds on the Loess Plateau, where integrated control measures reduced soil loss by an average of 49.6% (Liu, 1999). For individual storm events, 117 runoff-generating storms were recorded. In YDG, these storms generated an average runoff depth of 3.64 mm and a peak discharge of 0.78 m^3/s. In contrast, the same storms in CCZG resulted in an average runoff depth of 1.89 mm and a peak discharge of 0.28 m^3/s, reflecting reductions of 48% in runoff depth and 58.7% in peak discharge. The study found that while the measures effectively reduced runoff during individual storm events, their ability to mitigate peak discharge diminished as storm frequency increased. Specifically, runoff reduction effectiveness for storm recurrence intervals of less than 1, 1–2, 2–5, 5–10, 10–20, and over 20 years was 53.8%, 64.1%, 34.7%, 64.9%, 47.6%, and 50.7%, respectively. Corresponding reductions in peak flow discharge were 69.3%, 70.6%, 50.6%, 72.8%, 58.4%, and 15.8%. Additionally, soil loss was reduced with effectiveness rates of 61.8%, 55.4%, 46.7%, 69.2%, 57.2%, and 66.8% for the same intervals (Zhu, 2016).

5.4 Limitations of the Previous Research and Recommendations for Future Studies

Most previous studies evaluating the effectiveness of conservation measures on soil erosion have been conducted at the plot level, primarily focusing on sheet and rill erosion. As a result, these studies often overestimate the effectiveness of conservation measures when applied to larger watershed scales, as other types of erosion, such as gully erosion, mass movement, and tunnel erosion, also occur. Furthermore, many field studies have been conducted over relatively short periods, and due to varying rainfall conditions across different studies, drawing conclusive results is challenging, leading to conflicting findings. Therefore, it is essential to conduct long-term studies across various subregions, with monitoring not only at the plot level but also at the subcatchment level, to assess the effectiveness of conservation measures in reducing runoff and sediment under a wide range of rainfall conditions.

Research on the impact of revegetation on soil erosion often focuses on a single species. There have been a limited of studies examining the effects of mixed species, different types of vegetation, and the layered structures of various vegetation types. Besides, research on the role of biological crusts in soil conservation is limited. Few studies have investigated the effectiveness of vegetation in controlling mass movement, and the results from those studies are sometimes controversial. Therefore, further research in this area is needed.

As more sloped croplands are converted into terraces and vegetation land, erosion in gully areas becomes increasingly important. Future research should focus on how various conservation measures, such as check dams, natural revegetation, and artificial planting, can reduce soil loss in the gully areas. More experimental watersheds should be selected to monitor and evaluate the effectiveness of comprehensive conservation measures. Those small watershed monitoring programs can be linked to university extension services.

There are limited studies on the effectiveness of soil erosion control measures in the wind-water crisscrossed erosion region and the Pisha sandstone area of the Loess Plateau. Erosion control in these areas is challenging, and more research is needed to develop effective solutions for the region.

For soil erosion control to be sustainable, it must be closely linked to economic benefits, investment costs, and the motivation of local farmers. While most studies focus on the effectiveness of erosion control, few have addressed the integration of costs, benefits, and farmers' willingness to adopt these measures. This aspect should be further explored.

Finally, the impact of soil erosion control in the Loess Plateau region on the lower reaches of the Yellow River needs to be examined comprehensively. This includes exploring the connections between soil loss reduction in the Loess Plateau and sedimentation in the lower Yellow River channel, the development of the Yellow River delta, the impact of soil conservation on the decrease of streamflow in the lower river channel, and the relationship between a decrease in streamflow discharge and the sedimentation processes in the lower river channel.

References

Bai, Z. G. (1997). The loss reducing benefit of woods and grass in the flood of "94. 8. 4"in Wuding watershed. *Journal of Soil and Water Conservation in China, 7*, 17–19. (In Chinese).

Bai, H. Y., Tang, K. L., & Zhang, K. L. (1993). Study on man-made accelerated erosion of grassland being reclaimed by simulated rainfall experiment. *Memoir of Northwestern Institute of Soil and Water Conservation, Academia Sinica and Ministry of Water Resources, 17*, 87–93. (in Chinese).

Belnap, J., Weber, B., & Büdel, B. (2016). *Biological soil crusts as an organizing principle in drylands* (pp. 3–13). Springer International Publishing.

Cai, Q. G. (2001). Soil erosion and management on the loess plateau. *Journal of Geographical Sciences, 11*, 53–70.

Cai, Q. G., Chen, H., Luk, S. H., Jia, Z. J., Li, J. Y., & Chen, Y. Z. (1990). The formation of plow pan and its effect on soil and water loss as well as soil firtility. In Y. Z. Chen (Ed.), *Sino-Canadian symposium on experimental research on the law of soil erosion in the loess plateau region, Western Shanxi* (p. 141). Water Conservency and Electric Power Press. (in Chinese).

Cao, S., Chen, L., Liu, Z., & Wang, G. (2008). A new tree-planting technique to improve tree survival and growth on steep and arid land in the loess plateau of China. *Journal of Arid Environments, 72*(7), 1374–1382.

Chen, L., Wei, W., Fu, B., & Lü, Y. (2007). Soil and water conservation on the loess plateau in China: Review and perspective. *Progress in Physical Geography, 31*(4), 389–403.

Chen, L., Wang, J., Wei, W., Fu, B., & Wu, D. (2010). Effects of landscape restoration on soil water storage and water use in the loess plateau region, China. *Forest Ecology and Management, 259*(7), 1291–1298.

Chen, H., Zhang, X., Abla, M., Lü, D., Yan, R., Ren, Q., Ren, Z., Yang, Y., Zhao, W., Lin, P., & Liu, B. (2018). Effects of vegetation and rainfall types on surface runoff and soil erosion on steep slopes on the loess plateau, China. *Catena, 170*, 141–149.

Deng, L., Wang, K. B., Chen, M. L., Shangguan, Z. P., & Sweeney, S. (2013). Soil organic carbon storage capacity positively related to forest succession on the loess plateau, China. *Catena, 110*, 1–7.

Fang, N., Zeng, Y., Ran, L., Wang, Z., Lu, X., Wang, Z., Yang, X., Jian, J., Yu, Q., Ni, L., & Liu, C. (2023). Substantial role of check dams in sediment trapping and carbon sequestration on the Chinese loess plateau. *Communications Earth & Environment, 4*(1), 65.

Feng, X., Fu, B., Piao, S., Wang, S., Ciais, P., Zeng, Z., Lü, Y., Zeng, Y., Li, Y., Jiang, X., & Wu, B. (2016). Revegetation in China's loess plateau is approaching sustainable water resource limits. *Nature Climate Change, 6*(11), 1019–1022.

Fu, S., Liu, B., Zhang, G., Lu, B., & Ye, Z. (2010). Fish-scale pits reduce run-off and sediment. *Transactions of the ASABE, 53*, 157–162.

Fu, B., Liu, Y., Lü, Y., He, C., Zeng, Y., & Wu, B. (2011). Assessing the soil erosion control service of ecosystems change in the loess plateau of China. *Ecological Complexity, 8*(4), 284–293.

Gao, L., Bowker, M. A., Xu, M., Sun, H., Tuo, D., & Zhao, Y. (2017). Biological soil crusts decrease erodibility by modifying inherent soil properties on the loess plateau, China. *Soil Biology and Biochemistry, 105*, 49–58.

Guo, H., Zhang, B., Hill, R. L., Wu, S., Dong, Q., Sun, L., & Zhang, K. (2019). Fish-scale pit effects on erosion and water runoff dynamics when positioned on a soil slope in the loess plateau region, China. *Land Degradation & Development, 30*(15), 1813–1827.

Guo, W. Z., Wang, W. L., Xu, Q., Hu, J. J., & Zhu, L. L. (2021). Distribution, failure risk and reinforcement necessity of check-dams on the loess plateau: A review. *Journal of Mountain Science, 18*(2), 499–509.

Hamilton, H., & Luk, S. H. (1992). Nutrient loss in a small watershed: Examination of sources and transfer mechanisms. In F. T. Wang (Ed.), *Soil erosion management and application of GIS in the loess plateau region, Western Shanxi*. Science press. (in Chinese).

He, J., Wang, Q., Li, H., Tullberg, J. N., McHugh, A. D., Bai, Y., Zhang, X., McLaughlin, N., & Gao, H. (2009). Soil physical properties and infiltration after long-term no-tillage and ploughing on the Chinese loess plateau. *New Zealand Journal of Crop and Horticultural Science, 37*(3), 157–166.

Huang, M., Shao, M., Zhang, L., & Li, Y. (2003). Water use efficiency and sustainability of different long-term crop rotation systems in the loess plateau of China. *Soil and Tillage Research, 72*(1), 95–104.

Jiao, J. Y., Wang, W. Z., & Li, J. (1999). Analysis on soil and water conservation benefit of level terrace under different rainfall condition in loess hilly region. *Journal of Soil Erosion Soil Water Conservation, 5*, 59–63. (In Chinese with English abstract).

Jin, Z., Cui, B., Song, Y., Shi, W., Wang, K., Wang, Y., & Liang, J. (2012). How many check dams do we need to build on the loess plateau? *Environmental Science & Technology, 46*, 8527.

Li, F. R., Gao, C. Y., Zhao, H. L., & Li, X. Y. (2002). Soil conservation effectiveness and energy efficiency of alternative rotations and continuous wheat cropping in the loess plateau of Northwest China. *Agriculture, Ecosystems & Environment, 91*(1–3), 101–111.

Li, M., Yao, W., & Li, Z. (2005). Progress of the effect of grassland vegetation for conserving soil and water on loess plateau. *Advances in Earth Science, 20*(1), 74.

Li, P., Mu, X., Holden, J., Wu, Y., Irvine, B., Wang, F., Gao, P., Zhao, G., & Sun, W. (2017). Comparison of soil erosion models used to study the Chinese loess plateau. *Earth-Science Reviews, 170*, 17–30.

Liu, G. B. (1999). Soil conservation and sustainable agriculture on the loess plateau: Challenges and prospects. *Ambio, 28*, 663–668.

Liu, X., He, B., Li, Z., Zhang, J., Wang, L., & Wang, Z. (2011). Influence of land terracing on agricultural and ecological environment in the loess plateau regions of China. *Environmental Earth Sciences, 62*, 797–807.

Liu, Y., Zhao, W., Zhang, X., & Fang, X. (2016). Soil water storage changes within deep profiles under introduced shrubs during the growing season: Evidence from semiarid loess plateau, China. *Water, 8*(10), 475.

Liu, Y. F., Liu, Y., Shi, Z. H., López-Vicente, M., & Wu, G. L. (2020). Effectiveness of re-vegetated forest and grassland on soil erosion control in the semi-arid loess plateau. *Catena, 195*, 104787.

Liu, Y. F., Meng, L. C., Huang, Z., Shi, Z. H., & Wu, G. L. (2022). Contribution of fine roots mechanical property of Poaceae grasses to soil erosion resistance on the loess plateau. *Geoderma, 426*, 116122.

Lu, Y., Li, X., Xin, L., Song, H., & Wang, X. (2023). Mapping the terraces on the loess plateau based on a deep learning-based model at 1.89 m resolution. *Scientific Data, 10*(1), 115.

Shao, R., Zhang, B., Su, T., Long, B., Cheng, L., Xue, Y., & Yang, W. (2019). Estimating the increase in regional evaporative water consumption as a result of vegetation restoration over the loess plateau, China. *Journal of Geophysical Research: Atmospheres, 124*(22), 11783–11802.

Sun, W., Shao, Q., Liu, J., & Zhai, J. (2014). Assessing the effects of land use and topography on soil erosion on the loess plateau in China. *Catena, 121*, 151–163.

Wang, Y. K., Liu, B. Y., & Jiao, J. Y. (1992). Perspective of recovering the vegetation and cover percentage of forest and grass on the loess plateau. *Bulletin of Soil and Water Conservation, 12*(2), 55–60. (In Chinese).

Wang, Q., Bai, Y., Gao, H., He, J., Chen, H., Chesney, R. C., Kuhn, N. J., & Li, H. (2008). Soil chemical properties and microbial biomass after 16 years of no-tillage farming on the loess plateau, China. *Geoderma, 144*(3–4), 502–508.

Wang, Z., Liu, B., Liu, G., & Zhang, Y. (2009). Soil water depletion depth by planted vegetation on the loess plateau. *Science in China Series D: Earth Sciences, 52*(6), 835–842.

Wang, Y., Fu, B., Chen, L., Lu, Y., & Gao, Y. (2011). Check dam in the loess plateau of China: Engineering for environmental services and food security. *Environmental Science & Technology, 45*, 10299.

Wang, Z., Liu, G. B., Xu, M. X., Zhang, J., Wang, Y., & Tang, L. (2012). Temporal and spatial variations in soil organic carbon sequestration following revegetation in the hilly loess plateau, China. *Catena, 99*, 26–33.

Wang, S., Fu, B., Gao, G., Liu, Y., & Zhou, J. (2013). Responses of soil moisture in different land cover types to rainfall events in a re-vegetation catchment area of the loess plateau, China. *Catena, 101*, 122–128.

Wang, Y., Chen, L., Gao, Y., Wang, S., Lü, Y., & Fu, B. (2014). Carbon sequestration function of check-dams: A case study of the loess plateau in China. *Ambio, 43*, 926–931.

Wang, S., Fu, B., Piao, S., et al. (2015). Reduced sediment transport in the Yellow River due to anthropogenic changes. *Nature Geoscience, 9*, 38–41.

Wang, J., Fu, B., Lu, N., Wang, S., & Zhang, L. (2019). Water use characteristics of native and exotic shrub species in the semi-arid loess plateau using an isotope technique. *Agriculture, Ecosystems & Environment, 276*, 55–63.

Wei, W., Chen, L., Fu, B., Huang, Z., Wu, D., & Gui, L. (2007). The effect of land uses and rainfall regimes on runoff and soil erosion in the semi-arid loess hilly area, China. *Journal of Hydrology, 335*(3–4), 247–258.

Xia, L., Song, X., Fu, N., Cui, S., Li, L., Li, H., & Li, Y. (2019). Effects of forest litter cover on hydrological response of hillslopes in the loess plateau of China. *Catena, 181*, 104076.

Xiao, L., Zhao, R., & Kuhn, N. J. (2019). Straw mulching is more important than no tillage in yield improvement on the Chinese loess plateau. *Soil and Tillage Research, 194*, 104314.

Xu, X. Z., Zhang, H. W., & Zhang, O. Y. (2004). Development of check-dam systems in gullies on the loess plateau, China. *Environmental Science & Policy, 7*(2), 79–86.

Xu, Y. D., Fu, B. J., & He, C. S. (2013). Assessing the hydrological effect of the check dams in the loess plateau, China, by model simulations. *Hydrology and Earth System Sciences, 17*(6), 2185–2193.

Xu, M., Zhang, J., Liu, G. B., & Yamanaka, N. (2014). Soil properties in natural grassland, Caragana korshinskii planted shrubland, and Robinia pseudoacacia planted forest in gullies on the hilly loess plateau, China. *Catena, 119*, 116–124.

Xu, X., Zhu, T., Zhang, H., & Gao, L. (2020). *Experimental erosion: Theory and practice of soil conservation experiments*. Springer Nature.

Xu, Y., Zhu, G., Wan, Q., Yong, L., Ma, H., Sun, Z., Zhang, Z., & Qiu, D. (2021). Effect of terrace construction on soil moisture in rain-fed farming area of loess plateau. *Journal of Hydrology: Regional Studies, 37*, 100889.

Yang, X., Zhang, X., Lv, D., Yin, S., Zhang, M., Zhu, Q., Yu, Q., & Liu, B. (2020). Remote sensing estimation of the soil erosion cover-management factor for China's loess plateau. *Land Degradation & Development, 31*(15), 1942–1955.

Zeng, Q., Lal, R., Chen, Y., & An, S. (2017). Soil, leaf and root ecological stoichiometry of Caragana korshinskii on the loess plateau of China in relation to plantation age. *PLoS One, 12*(1), e0168890.

Zhang, X., Zhao, W., Liu, Y., Fang, X., & Feng, Q. (2016). The relationships between grasslands and soil moisture on the loess plateau of China: A review. *Catena, 145*, 56–67.

Zhang, B., Tian, L., Yang, Y., & He, X. (2022). Revegetation does not decrease water yield in the loess plateau of China. *Geophysical Research Letters, 49*(9), e2022GL098025.

Zhao, G., Mu, X., Wen, Z., Wang, F., & Gao, P. (2013). Soil erosion, conservation, and eco-environment changes in the loess plateau of China. *Land Degradation & Development, 24*(5), 499–510.

Zhao, C., Jia, X., & Zhu, Y. (2017). Long-term temporal variations of soil water content under different vegetation types in the loess plateau, China. *Catena, 158*, 55–62.

Zheng, F. L. (2006). Effect of vegetation changes on soil erosion on the loess plateau. *Pedosphere, 16*(4), 420–427.

Zhu, T. X. (2002). Effects of soil crusting on soil moisture, runoff and erosion: Field observation. *International Journal of Sediment Research, 17*, 298–303.

Zhu, T. X. (2003). Tunnel development over a 12-year period in a semi-arid catchment of the loess plateau, China. *Earth Surface Processes and Landforms: The Journal of the British Geomorphological Research Group, 28*(5), 507–525.

Zhu, T. X. (2016). Effectiveness of conservation measures in reducing runoff and soil loss under different magnitude–frequency storms at plot and catchment scales in the semi-arid agricultural landscape. *Environmental Management, 57*, 671–682.

Zhu, T. X., & Zhu, A. X. (2014). Assessment of soil erosion and conservation on agricultural sloping lands using plot data in the semi-arid hilly loess region of China. *Journal of Hydrology: Regional Studies, 2*, 69–83.

Zhu, T. X., Cai, Q. G., & Zeng, B. Q. (1997). Runoff generation on a semi-arid agricultural catchment: Field and experimental studies. *Journal of Hydrology, 196*(1–4), 99–118.

Zhu, P., Zhang, G., Wang, H., Yang, H., Zhang, B., & Wang, L. (2021). Effectiveness of typical plant communities in controlling runoff and soil erosion on steep gully slopes on the loess plateau of China. *Journal of Hydrology, 602*, 126714.

Chapter 6
Water Resources Management

Contents

6.1 Introduction

Water is essential for human society, playing a critical role in our survival and well-being. It is used across various sectors, including agriculture, industry, and households. Effective water resources management is crucial for maintaining the sustainability of water use, ensuring it meets the needs of both current and future generations (Wang et al., 2017). This objective aligns with one of the 17 United Nations Sustainable Development Goals (SDGs) (SDG, 2021). Water management involves managing both the quantity and quality of water to support economic growth, social benefits, and environmental sustainability. Achieving this goal requires collaboration and coordination among multiple stakeholders to ensure efficiency and long-term sustainability. Additionally, the variability of water availability over time and across regions adds complexity to this challenge. Furthermore, climate change introduces additional uncertainties into water resource management. A watershed serves as a natural unit that controls the movement of water, making it an ideal scale for water resources management. Managing water resources at the watershed level ensures that coordination between upstream and downstream areas is maintained, promoting effective and integrated water management throughout the entire watershed.

T. Zhu, *Watershed Environmental Degradation and Sustainable Development*, https://doi.org/10.1007/978-3-032-02448-0_6

Water is among the most important natural resources for economic development and environmental sustainability in the Yellow River basin. The region, characterized by arid, semi-arid, and semi-humid climates, faces water scarcity as the primary constraint to regional development. Home to approximately 110 million people, the basin includes 120 million hectares of farmland, representing 8% of China's total population and 17% of its agricultural land. However, water availability is highly limited, with an average of about 500 m^3 per capita per year—roughly a quarter of the national average (Chen et al., 2020). As agricultural, industrial, and urban development progress, concerns over water contamination are increasing as well. This chapter will focus on water distribution and management, addressing both the quantity and quality of water in the basin.

6.2 Water Resources and Usage

Water resources and usage in a basin are influenced by both natural and human factors. Natural factors include precipitation and its variations over time, evaporation, soil and vegetation cover, and groundwater recharge. Human factors encompass population density, land use, urbanization, dams and reservoirs, deforestation, and revegetation efforts, among others.

Within the Yellow River basin, there is great spatial variability of annual precipitation, ranging from less than 200 mm in the northwestern to more than 700 mm in the southeastern. However, the mean annual evapotranspiration exceeds annual precipitation in the entire basin, ranging from 650 mm in the upper source region to 1250 mm in the northwest part and southeast part of the basin (Wang et al., 2012). Water supply is greatly impacted by seasonality and climate variability. On average, 60% to 70% of annual precipitation occurs in the wet season of June through September. The seasonality of precipitation affects the temporal changes of streamflow discharges.

Yu et al. (2011) analyzed the flow discharge data of the Yellow River from 1976 to 2009 and found that the average annual flow discharge is 207.47×10^8 m^3 per year, with a maximum of 496×10^8 m^3 per year in 1983 and a minimum of 18.6×10^8 m^3 per year in 1997. Zhang et al. (2009a, 2009b) analyzed the stream flow discharges at six gauges of the Yellow River over the period of 50 years. The results showed that the available water resources in the basin have generally decreased over time, with an increase frequency of low-flow regime of streamflow but a decrease of high-flow regime of streamflow. The water deficit is more serious in spring and autumn compared to summer and winter.

The flow contribution to the Yellow River varies significantly along its upper, middle, and lower reaches. The upper reaches extend from the river's source to Hekou in Inner Mongolia, covering 3472 km and draining an area of 428,000 km^2. This section can be further divided into the source region and the area downstream of it. The source region, located in the northeastern Tibet Plateau, drains above Tongnanhai hydrological station, with a drainage area of about 121,972 km^2. which accounts for 16% of the total basin area, while contributing 35% of the total annual flow of the Yellow River (Lan et al., 2010; Meng et al., 2016). Flow levels remain high because of the low population density, which results in reduced demand from the agricultural, industrial, and domestic sectors. The river flows are contributed by rainfall, melting of snow cover, and thawing of permafrost soils and ground water. The mean annual precipitation is about 500 mm, ranging from 350 mm in the northwest to 700 mm in the southeast, among which more than 70% falls in the wet season from June to September (Meng et al., 2016). The duration of snow season is typically from November to April.

According to the 60-year (1950–2010) flow discharge recorded at the Tongnanhai station, flow discharge shows an overall slightly decline, though the decline between 1990–2010 is faster than 1950–1990. Meng et al. (2016) attributed the decline in river flows during the 1990s to both reduced precipitation and a slight increase in evaporation. However, Guan et al. (2021) argued that areal-average precipitation exhibited an insignificant decreasing trend from 1979 to 2016. Ma et al. (2021) showed that there is a significant increase in the precipitation in the source region of the Yellow River basin over the period from 1956 to 2018, but there was no significant change in runoff over this period. Since the beginning of this century, the temperature has risen rapidly in the source region. The rapid warming not only increases evaporation, but also contributed to the earlier spring peak flow and degradation of the permafrost ground. With rising temperature, the date of snowmelt occurs earlier over time. Guan et al. (2021) indicated that annual snow depth and number of snow days both showed a downward trend. The source region has an extensive permafrost distribution due to its high altitude, with a total area of about 25,000 km^2 (Jin et al., 2022). Sheng et al. (2020) reported that from 1972 to 2012, the area of degraded permafrost reached 1056 km^2 due to a rise in air temperatures beginning in the 1980s. By 2050, the degraded area could potentially double, and by 2100, it may expand even further to 3500 km^2, 10,000 km^2, or 18,492 km^2 under the RCP 2.6, RCP 6.0, and RCP 8.5 climate scenarios, respectively. Permafrost thawing contributes more to subsurface flow in rivers while reducing surface flow (Wang et al., 2018; Yang et al., 2023).

In the downstream area below the source region of the upper reaches, annual precipitation is significantly lower than in the source region, with rainfall averaging only about 150–250 mm, while potential evaporation is much higher. However, between Lanzhou and Hekou, a sharp population increase in cities like Lanzhou and Yinchuan significantly raises water demand. In addition, this section is home to two of the largest irrigation districts in the Yellow River basin: the Yingchuan and Hetao plains which withdraw substantial water from the river for agricultural irrigation. Annual irrigation water consumption was initially 5.581 billion m^3 for the Hetao Plain and about 6.54 billion m^3 for the Yingchuan Plain, though has reduced to approximately 4.38 billion m^3 and 4.369 billion m^3 per year, respectively, after the implementation of water-saving measures and policies (Xue & Ren, 2017).

The middle reaches of the Yellow River span from Hekou to Zhengzhou in Henan Province, stretching 1206 km and covering a drainage area of 344,000 km^2.This region, covering a major part of the Loess Plateau, has arid and semi-arid climates, with mean annual precipitation of about 500 mm, which gradually increases from the northwest to the southeast. More than 70% of the annual precipitation falls in the rainy season from May to September. Numerous studies have examined climate change in the region. While nearly all report a warming trend, findings on changes in precipitation patterns are less consistent. Peng et al. (2018) analyzed climate trends from 1901 to 2100 in the Loess Plateau region, identifying no significant trend in annual precipitation but a notable temperature increase, averaging 0.097 °C per decade. In contrast, Yao et al. (2005) and Xin et al. (2011) indicated a marked decline in annual precipitation over the past 50 years. Wang et al. (2012) reported a non-significant overall decline in precipitation (−29.11 mm over 50 years), with spatial variations: a significant drying trend in the northeast and a non-significant wetting trend in the northwest.

The agriculture land in the Loess Plateau is mainly rainfed. However, the flood plain areas adjacent to the tributaries of the Yellow River as well as some of the tablelands on the Loess Plateau still use in-stream water or groundwater for irrigation. For example, the Guanzhong Plain, located in Shaanxi Province, diverts water from the Wei River, the largest tributary of the Yellow River, to irrigate winter wheat. Stream flow discharge shows interannual and intraannual variability due to climate variability. The usage of the streamflows for irrigation is also limited by the high sediment concentration of river water due to the severe erosion in the region. In general, if the sediment concentration is more than 10% of the streamflow, it is not suitable for irrigation. The tablelands of the region often withdraw groundwater for crop irrigation. The overpumping ground water causes the rapid lowering of ground water table. For example, in some tableland areas, the ground water table decreased from a few meters below the surface before 1970s to more than 30 m in late 1990s. This region has a high population and also is home to important mining and energy industries, making water usage for domestic and industrial sectors critical. Water from the Yellow River and its tributaries support large cities like Xi'an and Taiyuan, as well as rural drinking water needs. Some of the large diversion engineering projects have been or are being completed, such as the Wanjiazhai Yellow River Diversion project in Shanxi, which has supplied more than 6 billion m^3 of water since its operation in 2003.

The lower reaches flow from Zhengzhou to the sea, with a length of 786 km and a drainage area of 23,000 km^2. This region has the highest annual precipitation in the basin, with a mean annual precipitation of over 700 mm. However, the drainage area is relatively small compared to the upper and middle reaches. This region is significantly impacted by irrigation activities in the North China Plain, as well as by economic development driven by high population densities in major cities like Zhengzhou, the capital of Henan Province. Due to extensive water usage upstream, the flow discharge at the Lijin gauge, near the river's mouth, is even lower than at the Huayuankou station and only slightly higher than at the Lanzhou gauge, which monitors a contributing area representing 30% of the total basin (Yang et al., 2004). Water storage in the reservoirs also greatly impacts the temporal pattern of streamflow discharges. There are some major reservoirs built across the main channel (e.g. Longyangxia, Liujiaxia, Sanmenxia and Xiaolongdi) and over 3000 water reservoirs in the basin, together with numerous check dams (Zhang et al., 2009a, 2009b). Since the 1970s, the lower stream of the Yellow River has experienced dry-up and the days of no flow peaked in 1997, with a total of 226 days (Xu et al., 2011). This severe situation has been improved in this century through regulations of several large reservoirs.

6.3 Water Quality Issues

Water quality in the Yellow River basin also faced serious challenges but has been improved substantially in recent years. The main pollutants are TN (total nitrogen),sediments, CODcr(dichromate oxidizability), NH_4^+-N(ammonia nitrogen), and TP(total phosphorous), lower dissolved oxygen (Tian et al., 2022; Quan et al., 2022). The contributing sources of the main pollutants are agriculture, mining industry, as well as domestic sewage.

Agricultural activities are mainly responsible for nitrate pollution in both surface and groundwater. NO_3^- pollution is observed in the Fen and Wei rivers on the Loess Plateau (Xiao et al., 2022). Hou et al. (2023) showed that, in the apple-producing areas of the Loess Plateau, chemical fertilizer was the highest contributor to NO_3^- in river water and groundwater, whereas sewage/manure was the most important nitrate source in tap water, whether irrigated or rainfed region. Nitrate pollution was more severe in irrigated than rain-fed regions in the region.

Application of fertilizers, agrochemicals and irrigation water, as well as atmospheric deposition also lead to heavy metals into soil and water. Xiao et al. (2019) conducted a comprehensive study on the sources, water quality and health risk of trace elements in the Loess Plateau. They found that The average trace elements concentrations were higher than the world average; Al, Fe, Li, B, As, and F had natural origins from loess weathering and leaching; Cr, Pb, Cd, Cu, Ag, and Tl were mainly from anthropogenic input; Co, Ni, and Mn were dominated by both anthropogenic and natural sources.

Coal is the dominant energy source in China and the Loess Plateau is one of the major coal mining areas in China. Coal mining can deteriorate water quality. Coal mining produced a large amount of solid wastes, such as coal gangues and tailings, which are buried or deposited in piles, which can have potential impacts on water quality of both surface and ground water in the region. Li et al. (2018) found that the groundwater in the mining area is generally of poor quality and is unsuitable for direct human consumption or irrigation. Na^+, SO_4^{2-}, Cl^-, F^-, TH, TDS, NO_3^-, and COD_{Mn} are the major pollutants responsible for the poor quality of the phreatic water, while Na^+, SO_4^{2-}, F^-, and TDS are the major constituents affecting the confined groundwater quality. Wang et al. (2016) reported water contamination in the Yuyang coal mine, indicating that the discharge of mine water could pose potential risk to groundwater quality, and protection measures of water quality in mining areas must be implemented. However, Wu et al. (2019) found that groundwater is generally suitable for irrigation in Yulin City of Shaaxi Province and that 47.73% of the collected samples are of excellent and good quality and are suitable for direct consumption.

In addition to coal mining, there are other metal mining activities in the Yellow River basin. The non-ferrous metal mining and smelting could produce serious pollution of heavy metals (Liao et al., 2006; Hu et al., 2014; Zang et al., 2017). Erosion and hydrological processes affect the redistribution of heavy metals in the watershed. Heavy metals are transported through runoff and sediment, subsequently accumulating in soils. This accumulation impacts agricultural products, ultimately posing health risks to humans (Zang et al., 2017).

The Ministry of Ecology and Environment of China, has established a water quality grading system known as the "Water Quality Index" (WQI) to assess and classify water quality in China. The WQI ranges from Grade I (excellent) to Grade V (extremely poor). In 2010, the mainstream of the Yellow River, with a total of evaluated river length of 3613.0 km, has the following distribution of water quality classes (Source: photoint.net):

- Class I ~ II: 31.7%
- Class III: 46.0%
- Class IV: 19.2%
- Class V: 3.1%

Since then, overall, water quality in the Yellow River basin shows a progressive improvement in recent years (Huang et al., 2021). In 2022, the Ministry of Ecology and Environment of China announced the entire main river channel had achieved Grade II status.

6.4 Discussion and Conclusions

Water availability in the Yellow River basin is generally limited due to the semi-arid and arid climates. Water management is further complicated by the great temporal and spatial variability in water resources. As economic development and improvement of people's living standards continue in the Yellow River basin, water consumption is rising accordingly. Xu et al. (2002) estimated that the water demands in the Yellow River basin were 50.9, 56.5, and 59.5 billion m^3 for 2010, 2020, and 2030 under the business-as-usual scenario. Without the development of alternative water sources, the water shortage is expected to reach 6.62 billion m^3 by 2030.

To ensure sustainable water management in the Yellow River Basin, a comprehensive approach is needed that focuses on both enhancing water supply and managing demand. A long-term solution to address water shortages could involve inter-basin water transfer projects, such as the South-North Water Transfer Project. The Eastern and Central routes indirectly benefit the lower reaches of the Yellow River by alleviating pressure on the river, which has been heavily over-exploited for agricultural, industrial, and municipal uses in the past. The Western Route, first proposed in the 1950s, aims to divert water from the upper reaches of the Yangtze

River and its tributaries—such as the Tongtian, Yalong, and Dadu rivers—to the upper Yellow River. The project envisions transferring approximately 17 billion cubic meters of water annually (New Security Beat, 2021). If implemented, this project would directly tackle the water scarcity issues in the basin. However, it faces significant technical, environmental, social, and economic challenges.

Increasing water use efficiency presents a significant opportunity to reduce water demand across agriculture, industry, and urban sectors. With agriculture accounting for over 60% of total water consumption in the basin (Wang et al., 2019; Liu et al., 2022), improving irrigation efficiency is essential. Although water consumption for irrigation in the Hetao and Yingchuan plains has decreased by roughly 30% due to the implementation of water policies and conservation measures, further progress is necessary. On the Loess Plateau, a range of measures in crop management should be employed to improve water efficiency through crop rotations and land use changes (Duan et al., 2024; Zhang et al., 2013, 2014, 2016). Regulating water through reservoirs improves supply reliability and ensures balanced distribution across the upper, middle, and lower reaches of the river. This regulation is crucial for flood mitigation, drought alleviation, and meeting ecological needs. A successful example of this is the elimination of dry-bed periods in the lower Yellow River since the 2000s. Finally, recycling treated wastewater, both from domestic and industrial sources, can serve as an effective way to supplement water supply, further supporting sustainable water management in the basin.

References

Chen, Y. P., Fu, B. J., Zhao, Y., Wang, K. B., Zhao, M. M., Ma, J. F., Wu, J. H., Xu, C., Liu, W. G., & Wang, H. (2020). Sustainable development in the Yellow River Basin: Issues and strategies. *Journal of Cleaner Production, 263*, 121223.

Duan, Y., Wang, W., Zhuo, L., Liu, Y., & Wu, P. (2024). Regional blue and green water-saving potential and regulation paths for crop production: A case study in the Yellow River Basin. *Agricultural Water Management, 291*, 108631.

Guan, X., Liu, C., Bao, Z., Jin, J., & Wang, G. (2021). Spatial-temporal variability of the snow over the Yellow River source region and its influencing climate factors. *China Environmental Science, 41*, 1045–1054.

Hou, L., Liu, Z., Zhai, B., Zhu, Y., & Xu, X. (2023). Contrasting water quality in response to long-term nitrogen fertilization in rainfed and irrigated apple-producing regions on China's loess plateau. *Agriculture, Ecosystems & Environment, 354*, 108561.

Hu, Y., Wang, D., Wei, L., & Song, B. (2014). Heavy metal contamination of urban topsoils in a typical region of loess plateau, China. *Journal of Soils and Sediments, 14*, 928–935.

Huang, J., Zhang, Y., Bing, H., Peng, J., Dong, F., Gao, J., & Arhonditsis, G. B. (2021). Characterizing the river water quality in China: Recent progress and on-going challenges. *Water Research, 201*, 117309.

Jin, X., Jin, H., Luo, D., Sheng, Y., Wu, Q., Wu, J., Wang, W., Huang, S., Li, X., Liang, S., & Wang, Q. (2022). Impacts of permafrost degradation on hydrology and vegetation in the source area of the Yellow River on northeastern Qinghai-Tibet plateau, Southwest China. *Frontiers in Earth Science, 10*, 845824.

Lan, Y., Zhao, G., Zhang, Y., Wen, J., Liu, J., & Hu, X. (2010). Response of runoff in the source region of the Yellow River to climate warming. *Quaternary International, 226*(1–2), 60–65.
Li, P., He, S., Yang, N., & Xiang, G. (2018). Groundwater quality assessment for domestic and agricultural purposes in Yan'an City, Northwest China: Implications to sustainable groundwater quality management on the loess plateau. *Environment and Earth Science, 77*, 775.
Liao, X. Y., Chen, T. B., Wu, B., Yan, X. L., Nie, C. J., Xie, H., Zhai, L. M., & Xiao, X. Y. (2006). Mining urban soil pollution: Concentrations and patterns of heavy metals in the soils of Jinchang, China. *Geographical Research, 25*(5), 843–852.
Liu, Y., Lin, Y., Huo, Z., Zhang, C., Wang, C., Xue, J., & Huang, G. (2022). Spatio-temporal variation of irrigation water requirements for wheat and maize in the Yellow River Basin, China, 1974–2017. *Agricultural Water Management, 262*, 107451.
Ma, L., Liu, Z., Zhao, B., Lyu, J., Zheng, F., Xu, W., & Gan, X. (2021). Variations of runoff and sediment and their response to human activities in the source region of the Yellow River, China. *Environmental Earth Sciences, 80*, 1–12.
Meng, F., Su, F., Yang, D., Tong, K., & Hao, Z. (2016). Impacts of recent climate change on the hydrology in the source region of the Yellow River basin. *Journal of Hydrology: Regional Studies, 6*, 66–81.
New Security Beat. (2021). https://www.newsecuritybeat.org/2021/08/build-build-western-route-chinas-south-north-water-diversion-project/
Peng, S., Gang, C., Cao, Y., & Chen, Y. (2018). Assessment of climate change trends over the loess plateau in China from 1901 to 2100. *International Journal of Climatology, 38*(5), 2250–2264.
Quan, J., Xu, Y., Ma, T., Wilson, J. P., Zhao, N., & Ni, Y. (2022). Improving surface water quality of the Yellow River Basin due to anthropogenic changes. *Science of the Total Environment, 836*, 155607.
SDG. (2021). *United Nations*. Department of Economic and Social Affairs Sustainable Development.
Sheng, Y., Ma, S., Cao, W., & Wu, J. (2020). Spatiotemporal changes of permafrost in the headwater area of the Yellow River under a changing climate. *Land Degradation & Development, 31*(1), 133–152.
Tian, Y., Wen, Z., Cheng, M., & Xu, M. (2022). Evaluating the water quality characteristics and tracing the pollutant sources in the Yellow River Basin, China. *Science of The Total Environment, 846*, 157389.
Wang, W., Shao, Q., Peng, S., Xing, W., Yang, T., Luo, Y., Yong, B., & Xu, J. (2012). Reference evapotranspiration change and the causes across the Yellow River Basin during 1957–2008 and their spatial and seasonal differences. *Water Resources Research, 48*(5), 1.
Wang, W., Qiang, Y., Wang, Y., Sun, Q., & Zhang, M. (2016). Impacts of Yuyang coal mine on groundwater quality in Hongshixia water source, Northwest China: A physicochemical and modeling research. *Exposure and Health, 8*, 431–442.
Wang, G., Zhang, J., Jin, J., Weinberg, J., Bao, Z., Liu, C., Liu, Y., Yan, X., Song, X., & Zhai, R. (2017). Impacts of climate change on water resources in the Yellow River basin and identification of global adaptation strategies. *Mitigation and Adaptation Strategies for Global Change, 22*, 67–83.
Wang, T., Yang, H., Yang, D., Qin, Y., & Wang, Y. (2018). Quantifying the streamflow response to frozen ground degradation in the source region of the Yellow River within the Budyko framework. *Journal of Hydrology, 558*, 301–313.
Wang, Y., Zhao, W., Wang, S., et al. (2019). Yellow River water rebalanced by human regulation. *Scientific Reports, 9*, 9707.
Wu, J., Zhou, H., He, S., & Zhang, Y. (2019). Comprehensive understanding of groundwater quality for domestic and agricultural purposes in terms of health risks in a coal mine area of the Ordos basin, north of the Chinese loess plateau. *Environmental Earth Sciences, 78*, 1–17.
Xiao, J., Wang, L., Deng, L., & Jin, Z. (2019). Characteristics, sources, water quality and health risk assessment of trace elements in river water and well water in the Chinese loess plateau. *Science of the Total Environment, 650*, 2004–2012.

Xiao, J., Lv, G., Chai, N., Hu, J., & Jin, Z. (2022). Hydrochemistry and source apportionment of boron, sulfate, and nitrate in the Fen River, a typical loess covered area in the eastern Chinese loess plateau. *Environmental Research, 206*, 112570.
Xin, Z., Yu, X., Li, Q., & Lu, X. X. (2011). Spatiotemporal variation in rainfall erosivity on the Chinese loess plateau during the period 1956–2008. *Regional Environmental Change, 11*, 149–159.
Xu, Z. X., Takeuchi, K., Ishidaira, H., & Zhang, X. W. (2002). Sustainability analysis for Yellow River water resources using the system dynamics approach. *Water Resources Management, 16*, 239–261.
Xu, X., Huang, G., Qu, Z., & Pereira, L. S. (2011). Using MODFLOW and GIS to assess changes in groundwater dynamics in response to water saving measures in irrigation districts of the upper Yellow River basin. *Water Resources Management, 25*, 2035–2059.
Xue, J., & Ren, L. (2017). Water productivity in the Hetao Irrigation District in Inner Mongolia by an agro-hydrological model. *Irrigation Science, 35*, 357–382.
Yang, D., Li, C., Hu, H., Lei, Z., Yang, S., Kusuda, T., Koike, T., & Musiake, K. (2004). Analysis of water resources variability in the Yellow River of China during the last half century using historical data. *Water Resources Research, 40*(6), 1.
Yang, J., Wang, T., Yang, D., & Yang, Y. (2023). Insights into runoff changes in the source region of Yellow River under frozen ground degradation. *Journal of Hydrology, 617*, 128892.
Yao, Y. B., Li, Y. H., Wang, Y. R., Zhang, X. Y., Li, Y. B., & Wei, F. (2005). Effects of the climate and climatic productivity in the loess plateau of China on global climate change. *Agricultural Research in the Arid Areas, 23*(2), 202–208.
Yu, J., Fu, Y., Li, Y., Han, G., Wang, Y., Zhou, D., Sun, W., Gao, Y., & Meixner, F. X. (2011). Effects of water discharge and sediment load on evolution of modern Yellow River Delta, China, over the period from 1976 to 2009. *Biogeosciences, 8*(9), 2427–2435.
Zang, F., Wang, S., Nan, Z., Ma, J., Zhang, Q., Chen, Y., & Li, Y. (2017). Accumulation, spatio-temporal distribution, and risk assessment of heavy metals in the soil-corn system around a polymetallic mining area from the loess plateau, Northwest China. *Geoderma, 305*, 188–196.
Zhang, Q., Xu, C. Y., & Yang, T. (2009a). Variability of water resource in the Yellow River basin of past 50 years, China. *Water Resources Management, 23*, 1157–1170.
Zhang, Q., Xu, C. Y., & Yang, T. (2009b). Variability of water resource in the Yellow River Basin of past 50 years, China. *Water Resources Management, 23*, 1157–1170.
Zhang, S., Sadras, V., Chen, X., & Zhang, F. (2013). Water use efficiency of dryland wheat in the loess plateau in response to soil and crop management. *Field Crops Research, 151*, 9–18.
Zhang, S., Sadras, V., Chen, X., & Zhang, F. (2014). Water use efficiency of dryland maize in the loess plateau of China in response to crop management. *Field Crops Research, 163*, 55–63.
Zhang, T., Peng, J., Liang, W., Yang, Y., & Liu, Y. (2016). Spatial–temporal patterns of water use efficiency and climate controls in China's loess plateau during 2000–2010. *Science of the Total Environment, 565*, 105–122.

Chapter 7
Biological Deprivation and Restoration

Contents

7.1 Natural Vegetation

The uplift of the Tibetan Plateau and Himalayas has had a profound impact on both global and regional climates, as well as monsoon intensity (Wang et al., 2008). This uplift has contributed to an overall drying trend in the Yellow River basin, influenced by the development of a strong Siberian-Mongolian High over northern Asia during winter (Li & Fang, 1999; Sun et al., 1997; Ding et al., 1997). Such climate changes have shaped the natural vegetation of the Yellow River basin, creating distinct spatial variations across its upper, middle, and lower reaches.

Many studies have explored the evolution of vegetation in response to the uplift of the Tibet-Qinghai Plateau (Tang et al., 2006; Li et al., 2009; Zhang et al., 2012; Wen et al., 2014; Favre et al., 2015; Johansson et al., 2007). Fossil pollen data from Tianchi Lake and LiuPanshan Mountain reveal changes in vegetation cover during the Holocene period over the past 6200 years. Initially, between 6200 and 2900 cal yr. BP, the area was covered by a mixed deciduous–coniferous forest dominated by Betula, Quercus, Picea, and Pinus. This transitioned to a steppe forest with both trees (Betula, Picea, and Pinus) and herbs (mainly Artemisia) from 2900 to 1100 cal yr. BP. After 1100 cal yr. BP, the landscape became more open, dominated by Artemisia and Poaceae (Zhao et al., 2010).

T. Zhu, *Watershed Environmental Degradation and Sustainable Development*,
https://doi.org/10.1007/978-3-032-02448-0_7

There are still differing perspectives on the natural vegetation cover of the Loess Plateau in the middle reaches. Shi (1988) and Shi et al. (1985) believe, based on historical documents, that the region was primarily forested. Liu et al. (1996) argue that forest-steppe was predominant on the Loess Plateau, while others, such as Xie et al. (2002) and Jiang et al. (2003), suggest that steppe vegetation was the dominant cover in the area. Many studies have shown that natural vegetation exhibits significant spatial variation across the diverse topographies of the Loess Plateau (Liu et al., 1996; Lü et al., 2003). The tableland surfaces did not develop dense, stable forest cover; instead, they were characterized by steppe and grass-shrub landscapes interspersed with scattered deciduous trees, as indicated by pollen preserved in loess deposits (Sun et al., 1997; Lü et al., 2003). Forests were mainly concentrated in rocky mountainous areas, river valleys, gullies, and along the southern edge of the Loess Plateau (Ju & Chen, 1998). Li et al. (2003) analyzed pollen records from the southern edge of the Loess Plateau and identified four episodes of vegetation change over the past 12,000 years: from steppe to humid-grassland, back to steppe, followed by humid-grassland, and finally to grassland. Jiang et al. (2013), using pollen samples from six loess profiles across the Loess Plateau, indicated that during dry-cold periods since the last Glacial Maximum, the northwest Loess Plateau supported desert steppe vegetation, while the southeastern region was characterized by steppe vegetation. Steppe vegetation prevailed in the northwest, whereas meadow steppe vegetation dominated in the southeastern Loess Plateau.

In the lower reaches of the Yellow River basin, frequent shifts in river channels and high sediment deposition created a lake-swamp-floodplain environment. This area featured intrazonal grasslands in the North China Plain's hinterland, while the surrounding hills and mountains were dominated by mixed broadleaf-coniferous forests during the mid-Holocene (Li & Liang, 1985; Kong et al., 1992; Zhang et al., 2022; Li et al., 2022).

In the Yellow River delta, natural vegetation has been shaped by the interaction of climate changes and depositional processes. It is widely believed that the delta began to form at least by the late Pleistocene (Zhang, 1989; Zhu, 1989). During warmer periods, such as the middle Holocene, rising sea levels led to transgression, and the delta was populated by aquatic plants, marine algae, herbs, and both evergreen and deciduous broadleaf trees. In contrast, during dry and cold episodes, the area was mainly covered by herbs and aquatic plants, indicating a delta-front environment driven by delta progradation (Yi et al., 2003).

7.2 Human Destruction of Vegetation

The current vegetation pattern in the Yellow River basin, reflecting the combined effects of climate change and human activities on vegetation cover in recent history. These impacts have varied over time and across different parts of the basin, influenced by factors such as land use changes, revegetation efforts, and the vulnerability of natural conditions, as well as climate variability.

In the Yellow River's source region, climate changes, expansion of dry farmland, reduction of wetlands, overgrazing on grasslands have all contributed to vegetation changes. Opinions on the impact of climate change on vegetation in this region remain divided. Some studies indicate that rising temperatures have degraded meadow and alpine grassland ecosystems, resulting in reduced vegetation cover, declining plant species populations, and a reversal of ecological succession (Zhou et al., 2005; Yang et al., 2010). Conversely, other research suggests that warmer temperatures due to climate change promote vegetation growth in cold regions (Wang et al., 2016). Additional studies highlight alternating periods of vegetation increase and decrease in response to climate changes; during times of rising temperatures and increased precipitation, vegetation cover expands, whereas in periods of rising temperatures coupled with decreased precipitation, vegetation cover declines (Guo et al., 2016; Wang et al., 2020). The primary human-driven factor contributing to grassland degradation in the region is prolonged overgrazing, which is often exacerbated by rodents' excessive burrowing and gnawing, further intensifying grassland degradation and desertification. Wetland drainage, including marshes and peatlands, began in the 1950s and accelerated in the 1970s with large-scale drainage efforts aimed at expanding grassland and farmland. Converting wetlands to grasslands, which are then overgrazed, has led to reduced land productivity and further desertification (Zeng et al., 2003). However, there remains significant uncertainty in quantitatively assessing the contribution of each factor to grassland degradation, along with a limited understanding of how economic policies could effectively support grassland sustainability in the region (Harris, 2010).

In the upper and middle reaches of the Yellow River basin below the source region, vegetation destruction by humans has been extremely severe due to population growth pressures, especially on the Loess Plateau region. Archaeological evidence reveals that human settlement in the Loess Plateau dates back at least 8000 years. Population estimates indicate around ten million people during the Western Han Dynasty (2 A.D.), 11.5 million in the mid-Ming Dynasty (1491 A.D.), and over 100 million people today (Lu et al., 1996). The main causes of vegetation destruction are clearing natural vegetation and expanding cultivated farmlands. Wang et al. (2006) estimated that forest coverage declined from 53% during the Western Zhou Dynasty (1066–771 B.C.) to 9% by 1998. In the Ziwuling region of Shanxi Province, Zheng (2006) indicates much of the secondary forest has been extensively cleared due to human activities in modern times. Natural vegetation in the tableland areas (Yuan) was completely removed and converted to farmland, with similar patterns observed in the hilly and gully regions. Human activities have also significantly impacted plant diversity in the region. The expansion of cropland, overgrazing, desertification, mining, and other human pressures threaten plant diversity. Currently, over 3000 species of vascular plants grow in the Loess Plateau, including around 100 endangered species, such as indigenous Glycine soja, Paeonia suffruticosa var. spontanea, Astragalus membranaceus, and Xanthoceras sorbifolia (Fu, 1989, 1992, 2000; Jiang et al., 2013). Vegetation destruction has also led to plateau dissection and land degradation, further hindering vegetation growth. Observations indicate that steep gully sideslopes in watersheds undergoing natural restoration in recent decades in the hilly-gully region of the Loess Plateau still exhibit overall poor vegetation coverage. Fresh scars of mass movement, along with very low organic matter in weathered loess deposits on cliffs, hinder plant rooting. This issue is even more pronounced in the Pisha sandstone region, covering 16,700 km^2 in the northern Loess Plateau. Composed of thick layers of sandstone, arenaceous shale, and poorly bonded mudstone, Pisha sandstone is extremely hard when dry but rapidly disintegrates when wet (Chen, 2012; Ma & Zhang, 2016; Liang et al., 2019). Once surface soils are eroded by wind and water, the Pisha sandstone becomes exposed. Due to its low nutrient content and high erosion rates, this area has an extremely low vegetation survival rate and sparse coverage.

The desertification area in the middle reaches of the Yellow River basin is primarily located in the transitional zone between the Maowusu Sandy Area and the Loess Plateau, resulting from vulnerable natural conditions and human activities. This region experiences low precipitation, high potential evaporation, frequent strong winds, and abundant sand sources. The primary sand sources contributing to desertification originated from Quaternary aeolian sands (Dong et al., 1988; Zhang et al., 2003). Before the 1980s, human activities such as overgrazing in sparsely vegetated areas, unrestricted woodcutting for firewood, and conversion of grassland to farmland dominated the destruction of biological resources, causing the sandified area to expand. Studies show that the rate of sandified land expansion in northern China was 1560 km^2 per year between the late 1950s and 1975, 2100 km^2 per year from 1975 to 1987, and 3600 km^2 per year from 1987 to 2000 (Wang et al., 2004; Karnieli et al., 2014).

The natural vegetation in the North China Plain, located in the lower reaches of the Yellow River basin, has been largely cleared for agricultural production. Vegetation in the limited riparian wetlands has also declined due to dam construction, urbanization and tourism development (He et al., 2015). Vegetation in the Yellow River delta area faces threats from climate change, soil salinization, urbanization, and agricultural and industrial development. Since vegetation in the delta depends directly on freshwater, reduced river flow to the Bohai Sea (Yu et al., 2011), along with oil extraction and urban expansion in Dongying, Shandong Province, has led to significant wetland loss. The total natural wetland area shrank rapidly from 2565.6 km^2 in 1986 to 1574.5 km^2 in 2008, a decrease of 991.1 km^2 (Wang et al., 2012), resulting in reduced vegetation cover and biodiversity.

7.3 Revegetation and Its Potential Impact

While the natural vegetation in the basin has been greatly destructed by human activities in the recent history, a large scale of revegetation has also been launched in the last few decades, through "Three-North Shield Belt" and "Grain for Green" programmes. Jian et al. (2024) indicated that over the period of 1980 to 2015, the forestland, mainly converted from steep cropland and grassland, has increased by 2592 km^2, and the grassland was increased by 3520 km^2 transferred from unused land.

The "Three-North" Shelterbelt Project (TNSP) was launched in 1978 to address ecological challenges across northeastern, northern, and northwestern China. Spanning an area of 4,069,000 km^2 and encompassing 13 provinces—six of which are in the Yellow River basin, including Shanxi, Shaanxi, Gansu, and Qinghai provinces, as well as the Inner Mongolia and Ningxia autonomous regions—the project aims to promote sustainable development in these fragile regions (Li et al., 2012). With a planned duration until 2050 and a total investment of CNY 93.3 billion, TNSP has made significant progress. Over four decades, reforestation efforts have covered 30.14 million hectares, increasing forest cover from 5.05% in 1978 to 13.57% (Zhai et al., 2023). The initiative has generated multiple benefits, such as controlling soil erosion, combating desertification, reducing dust storms, enhancing carbon sequestration, improving crop yields, and fostering economic growth (Yan et al., 2015; Jia et al., 2015; Chu et al., 2019; Cao et al., 2020; Liu et al., 2021; Zhang et al., 2021). For instance, between 2004 and 2014, the extremely severe and severe sandy areas in the Maowusu Sandy Area decreased by 331,000 ha and 595,000 ha, respectively. From 1978 to 2017, the project contributed an estimated 2310 Tg of carbon sequestration (Zhai et al., 2023). Soil erosion caused by wind and water has significantly decreased on the Loess Plateau, although it is challenging to determine the exact contribution of the TNSP to this reduction. However, the project faces sustainability concerns due to the semi-arid and arid climates. Issues such as limited water resources and challenges with tree survival and healthy growth (e.g., the prevalence of stunted, aged trees) have been noted (Wang et al., 2010).

The "Grain for Green" program, launched in 1999, is an ambitious revegetation initiative aimed at increasing forest cover and reducing soil erosion by converting steep agricultural lands into forestland or grassland (Zhou et al., 2012; Li et al., 2019). Recognized as the world's largest vegetation restoration project (Chen et al., 2015), it has had a significant impact on forest coverage. Fu et al. (2017) reported a 6.2% increase in forest cover between 2000 and 2010 due to the conversion of slope farmlands. The program promotes revegetation through plantation or natural recovery, with some grasslands transitioning into forests via secondary succession. On the Loess Plateau, over 30,000 km^2 of cropland has been converted into forestland or grassland, boosting vegetation coverage by 25% (Shi et al., 2020). Using satellite remote sensing data from MODIS (Moderate Resolution Imaging Spectroradiometer), Xiao (2014) estimated that forest coverage expanded by 18,900 km^2 from 2000 to 2010, increasing from 7.44% to 10.99%, representing a relative 41% rise in tree coverage. Satellite remote sensing has also been used to assess vegetation dynamics in the region. Li et al. (2019) observed that the growing season's mean NDVI (normalized-difference vegetation index) and annual mean NDVI both increased between 1999 and 2015, with slopes of 4.75×10^{-3} and 2.32×10^{-3} year^{-1}, respectively. Although research consistently shows increased vegetation coverage since the program's inception, there remains no consensus on the precise percentage of the revegetation area achieved.

There are some concerns and debates about reforestation. Artificial revegetation forests often demand more soil water in this water resources limited region. It is important to understand the soil water carrying capacity for vegetation in the critical zone of the Loess Plateau (Shao et al., 2018). Large scale afforestation can deplete soil moisture and form the deep "dry soil layer", particularly in the drier parts of the Loess Plateau, leading to low survival rates of the trees, or slowly growing, so called "little old man tree". Chen et al. (2015) argue that continuing expansion of revegetation could cause more harm than good to environment and local communities. Conventional afforestation is usually lack of species diversity and some of the exotic species are not suitable to the area. Research indicates that revegetation can significantly influence hydrological processes. Chang and Lei (2022) found that intensive revegetation in the Dali River basin, a sub-watershed of the Yellow River basin, increased the Normalized Difference Vegetation Index (NDVI) by 6.9 between 2000 and 2010, accounting for 29.2% of the total reduction in runoff. In contrast, Yu et al. (2023) argued that while revegetation does reduce runoff in the Jing River basin, another sub-watershed of the Yellow River basin, the impact is primarily driven by natural vegetation (mainly grassland) rather than artificial revegetation efforts.

In addition to the above large scale revegetation projects in the middle reaches of the Yellow River, different measures have been taken to restore vegetation in the source region of the Yellow River basin, referring to the area above Tangnaihai, occupying 16% of the whole basin area (Guo et al., 2021). The measures include the establishment of nature reserve areas, improvements of grazing management, and conversion of crop land to vegetation land. Nevertheless, Rong and Long (2021) indicate that those measures have not increased vegetation cover significantly due to the harsh natural conditions where the climate is dominated by the alpine continental climates characterized by cold and dry weather.

7.4 Discussion and Conclusions

Researchers have used various methods, such as pollen analysis and historical documentation, to reconstruct the history of the natural vegetation and its temporal changes. However, a consensus remains elusive, particularly regarding the Loess Plateau. This lack of agreement stems from challenges such as limited pollen sampling sites, poor pollen preservation in loess deposits. Historical documents from the Loess Plateau region have been used to reconstruct the history of vegetation changes, but this approach may involve significant uncertainty and potential errors due to vague, non-quantitative, and inconsistent documentation. The vegetation types and coverage described in historical records are often site-specific, influenced by factors such as topography and microclimatic conditions. Extrapolating this information to a wider region may introduce significant uncertainties and errors.

The Yellow River basin has undergone significant vegetation destruction due to rapid population growth and inappropriate land use in recent history. Large-scale vegetation restoration programs, such as the Grain for Green Program, have been launched in recent decades to increase vegetation cover. While extensive studies have assessed revegetation efforts, there is considerable disagreement on several aspects. These include the extent of vegetation area increase, the proportion of increase in forest versus grassland, the temporal changes in vegetation cover, and the roles of climate variability (particularly precipitation) and human planting. Additionally, questions remain regarding the most effective methods for vegetation recovery, how the spatial variability of climate conditions should be factored into revegetation efforts, and how human planting and natural vegetation recovery can effectively work together to maximize the benefits of the revegetation efforts.

Studies on the impact of revegetation on the environment have produced varying results. The role of revegetation in reducing soil loss and the influence of changes in precipitation on soil erosion over the past few decades, as well as the contributions of revegetation, check dams, and terrace construction, remain topics of debate. Additionally, the effects of revegetation on hydrological cycles, particularly concerning deep soil water content and groundwater, are still uncertain. Recent studies have indicated a decline in deep soil water content, raising concerns about the long-term sustainability of reforestation efforts (Shao et al., 2018; Shao et al., 2019). Ongoing monitoring is crucial to better understand the water consumption patterns of restored vegetation and to ensure the sustainable management of water resources. Models have been used to assess the impact on evapotranspiration, but the results vary, and whether large-scale revegetation can increase precipitation remains a matter of debate.

Future climate change presents further challenges to the sustainability of artificially planted forests. Rising temperatures are expected to increase evaporation, worsening water scarcity. The region's semi-arid and arid climates, which are already susceptible to drought, are particularly vulnerable to these changes. This highlights the importance of continued research and collaborative efforts to improve vegetation restoration strategies.

Revegetation efforts must strike a balance between ecological restoration, crop production, economic development, and long-term sustainability. It is essential to maintain base crop fields that can produce sufficient food for local farmers, while also incorporating revegetation into the economic benefits of local communities. Current reforestation projects are largely funded by the government, but for long-term success, it is crucial to actively involve local farmers in the management of artificial forests. A comprehensive approach that combines ecological restoration with community participation is vital for optimizing vegetation recovery and sustainability in the Yellow River basin.

References

Cao, S., Suo, X., & Xia, C. (2020). Payoff from afforestation under the three-north shelter forest program. *Journal of Cleaner Production, 256*, 120461.

Chang, Y., & Lei, H. (2022). How intensive revegetation affects runoff in semiarid watersheds: A case study based on an integrated modelling framework. *Ecohydrology, 15*(7), e2451.

Chen, W. D. (2012). The Pisha sandstone in Yellow River basin-nonfarming land and watch as picture scroll. *National Geography of China, 5*, 87–93.

Chen, Y., Wang, K., Lin, Y., Shi, W., Song, Y., & He, X. (2015). Balancing green and grain trade. *Nature Geoscience, 8*, 739–741. https://doi.org/10.1038/ngeo2544

Chu, X., Zhan, J., Li, Z., Zhang, F., & Qi, W. (2019). Assessment on forest carbon sequestration in the three-north shelterbelt program region, China. *Journal of Cleaner Production, 215*, 382–389.

Ding, Z. L., Sun, J. M., Zhu, R. X., et al. (1997). Eolian origin of the red clay in the loess plateau and implications for Pliocene climatic changes. *Quaternary Sciences, 17*, 147.

Dong, G., Gao, S., & Jin, J. (1988). The formation, evolvement and causes of MU US Desert. *Science China, B6*, 633–642.

Favre, A., Päckert, M., Pauls, S. U., Jähnig, S. C., Uhl, D., Michalak, I., & Muellner-Riehl, A. N. (2015). The role of the uplift of the Qinghai-Tibetan plateau for the evolution of Tibetan biotas. *Biological Reviews, 90*(1), 236–253.

Fu, K. J. (1989). *The Flora of loess plateau V*. Science and Technology Literature Press.

Fu, K. J. (1992). *The Flora of loess plateau* (Vol. II). China Forestry Press.

Fu, K. J. (2000). *The Flora of loess plateau* (Vol. I). Science Press.

Fu, B., Wang, S., Liu, Y., Liu, J., Liang, W., & Miao, C. (2017). Hydrogeomorphic ecosystem responses to natural and anthropogenic changes in the loess plateau of China. *Annual Review of Earth and Planetary Sciences, 45*, 223–243.

Guo, B., Zhou, Y., Zhu, J. F., Liu, W. L., Wang, L. T., Yan, F. L., Wang, F., Yang, G., Luo, W., & Jiang, L. (2016). Spatial patterns of ecosystem vulnerability changes during 2001–2011 in the three-river source region of the Qinghai-Tibetan plateau, China. *Journal of Arid Land, 8*, 23–35.

Guo, B., Niu, Y., Mantravadi, V. S., Zhang, L., & Liu, G. (2021). The variation of rainfall runoff after vegetation restoration in upper reaches of the Yellow River by the remote sensing technology. *Environmental Science and Pollution Research, 28*, 50707–50717.

Harris, R. B. (2010). Rangeland degradation on the Qinghai-Tibetan plateau: A review of the evidence of its magnitude and causes. *Journal of Arid Environments, 74*(1), 1–12.

He, X., Hörmann, G., Strehmel, A., Guo, H., & Fohrer, N. (2015). Natural and anthropogenic causes of vegetation changes in riparian wetlands along the lower reaches of the Yellow River, China. *Wetlands, 35*, 391–399.

Jia, K., Liang, S. L., Liu, J. Y., Li, Q. Z., Wei, X. Q., Yuan, W. P., & Yao, Y. J. (2015). Forest cover changes in the three-north shelter forest region of China during 1990 to 2005. *Journal of Environmental Informatics, 26*(2), 112–120.

Jian, S., Wang, A., Hu, C., & Yan, D. (2024). Effect of landscape restoration on evapotranspiration and water use in the Yellow River Basin, China. *Acta Geophysica, 72*(1), 341–356.

Jiang, Y., Kang, M., Gao, Q., He, L., Xiong, M., Jia, Z., & Jin, Z. (2003). Impact of land use on plant biodiversity and measures for biodiversity conservation in the loess plateau in China–a case study in a hilly-gully region of the northern loess plateau. *Biodiversity and Conservation, 12*, 2121–2133.

Jiang, W., Cheng, Y., Yang, X., & Yang, S. (2013). Chinese loess plateau vegetation since the last glacial maximum and its implications for vegetation restoration. *Journal of Applied Ecology, 50*(2), 440–448.

Johansson, U. S., Alström, P., Olsson, U., Ericson, P. G., Sundberg, P., & Price, T. D. (2007). Build-up of the Himalayan avifauna through immigration: A biogeographical analysis of the Phylloscopus and Seicercus warblers. *Evolution, 61*(2), 324–333.

Ju, T. Z., & Chen, X. L. (1998). Research of vegetational succession on the Xiaolongshan Mt. since middle Holocene. *Acta Botanica Boreali-Occidentalia Sinica (in Chinese), 18*(2), 292–299.

Karnieli, A., Qin, Z., Wu, B., Panov, N., & Yan, F. (2014). Spatio-temporal dynamics of land-use and land-cover in the mu us sandy land, China, using the change vector analysis technique. *Remote Sensing, 6*(10), 9316–9339.

Kong, Z. C., Du, N. Q., Xu, Q. H., & Tong, G. B. (1992). Paleoclimatic fluctuations reflected in flora of Holocene Megathermal in the northern part of China. In *The climates and environments of Holocene megathermal in China* (pp. 48–65).

Li, J., & Fang, X. (1999). Uplift of the Tibetan plateau and environmental changes. *Chinese Science Bulletin, 44*, 2117–2124.

Li, W. Y., & Liang, Y. L. (1985). Vegetation and environment of the hypsithermal interval of Holocene in eastern Hebei plain. *Journal of Integrative Plant Biology, 27*(6), 1.

Li, X., Zhou, J., & Dodson, J. (2003). The vegetation characteristics of the 'Yuan' area at Yaoxian on the loess plateau in China over the last 12 000 years. *Review of Palaeobotany and Palynology, 124*(1–2), 1–7.

Li, J., He, Q., Hua, X., Zhou, J., Xu, H., Chen, J., & Fu, C. (2009). Climate and history explain the species richness peak at mid-elevation for Schizothorax fishes (Cypriniformes: Cyprinidae) distributed in the Tibetan plateau and its adjacent regions. *Global Ecology and Biogeography, 18*(2), 264–272.

Li, M. M., Liu, A. T., Zou, C. J., Xu, W. D., Shimizu, H., & Wang, K. Y. (2012). An overview of the "three-north" shelterbelt project in China. *Forestry Studies in China, 14*, 70–79.

Li, G., Sun, S., Han, J., Yan, J., Liu, W., Wei, Y., Lu, N., & Sun, Y. (2019). Impacts of Chinese grain for green program and climate change on vegetation in the loess plateau during 1982–2015. *Science of the Total Environment, 660*, 177–187.

Li, B., Wang, S., Li, Y., Xu, Q., Ge, Y., Fan, B., Zhang, S., Zhang, N., & Xie, M. (2022). Quantitative vegetation reconstruction in the central North China plain of the last 3000 years based on the REVEALS model. *Quaternary International, 641*, 39–50.

Liang, Z., Wu, Z., Yao, W., Noori, M., Yang, C., Xiao, P., Leng, Y., & Deng, L. (2019). Pisha sandstone: Causes, processes and erosion options for its control and prospects. *International Soil and Water Conservation Research, 7*(1), 1–8.

Liu, T. S., Guo, Z. T., Wu, N. Q., & Lu, H. Y. (1996). Prehistoric vegetation on the loess plateau: Steppe or forest? *Journal of Southeast Asian Earth Sciences, 13*(3–5), 341–346.

Liu, X., Liang, S., Li, B., Ma, H., & He, T. (2021). Mapping 30m fractional Forest cover over China's three-north region from Landsat-8 data using ensemble machine learning methods. *Remote Sensing, 13*, 2592.

Lu, J., Wang, J., Liu, X., & Sun, L. (1996). Studies on ecological limits to agricultural development on loess plateau: From the change of population capacity. *Ecological Economics, 4*, 44–47.

Lü, H., Liu, D., & Guo, Z. (2003). Natural vegetation of geological and historical periods in loess plateau. *Chinese Science Bulletin, 48*, 411–416.

Ma, W., & Zhang, X. (2016). Effect of Pisha sandstone on water infiltration of different soils on the Chinese loess plateau. *Journal of Arid Land, 8*, 331–340.

Rong, T., & Long, L. H. (2021, March). Quantitative assessment of NPP changes in the Yellow River source area from 2001 to 2017. *IOP Conference Series: Earth and Environmental Science, 687*(1), 012002.

Shao, M. A., Wang, Y., Xia, Y., & Jia, X. (2018). Soil drought and water carrying capacity for vegetation in the critical zone of the loess plateau: A review. *Vadose Zone Journal, 17*(1), 1–8.

Shao, R., Zhang, B., Su, T., Long, B., Cheng, L., Xue, Y., & Yang, W. (2019). Estimating the increase in regional evaporative water consumption as a result of vegetation restoration over the loess plateau, China. *Journal of Geophysical Research: Atmospheres, 124*(22), 11783–11802.

Shi, N. H. (1988). The research on the changing of Forest in history. *The Historical Geography Works of China, 1*, 1–17. (in Chinese).

Shi, N. H., Cao, E. Q., & Zhu, S. G. (1985). *The vicissitude of Forest and grass in loess plateau.* Shaanxi People's Press. (in Chinese).

Shi, P., Feng, Z., Gao, H., Li, P., Zhang, X., Zhu, T., Li, Z., Xu, G., Ren, Z., & Xiao, L. (2020). Has "grain for green" threaten food security on the loess plateau of China? *Ecosystem Health and Sustainability, 6*(1), 1709560.

Sun, X., Song, C., Wang, F., & Sun, M. (1997). Vegetation history of the loess plateau of China during the last 100,000 years based on pollen data. *Quaternary International, 37*, 25–36.

Tang, Z., Wang, Z., Zheng, C., & Fang, J. (2006). Biodiversity in China's mountains. *Frontiers in Ecology and the Environment, 4*(7), 347–352.

Wang,L., Shao, M.A., Wang,Q. and Gale.W.L., 2006. Historical changes in the environment of the Chinese Loess Plateau. *Environmental Science & Policy. 9*(7–8), 675–684.

Wang, T., Wu, W., & Xue, X. (2004). Spatial-temporal changes of sandy desertified land during last 5 decades in northern China. *Acta Geographica Sinica-Chinese Edition, 59*(2), 203–212.

Wang, C., Zhao, X., Liu, Z., Lippert, P. C., Graham, S. A., Coe, R. S., Yi, H., Zhu, L., Liu, S., & Li, Y. (2008). Constraints on the early uplift history of the Tibetan plateau. *Proceedings of the National Academy of Sciences, 105*(13), 4987–4992.

Wang, X. M., Zhang, C. X., Hasi, E., & Dong, Z. B. (2010). Has the three norths Forest shelterbelt program solved the desertification and dust storm problems in arid and semiarid China? *Journal of Arid Environments, 74*(1), 13–22.

Wang, M., Qi, S., & Zhang, X. (2012). Wetland loss and degradation in the Yellow River delta, Shandong Province of China. *Environmental Earth Sciences, 67*, 185–188.

Wang, X. Y., Yi, S. H., Wu, Q. B., Yang, K., & Ding, Y. J. (2016). The role of permafrost and soil water in distribution of alpine grassland and its NDVI dynamics on the Qinghai-Tibetan plateau. *Global and Planetary Change, 147*, 40–53.

Wang, R., Dong, Z., & Zhou, Z. (2020). Effect of decreasing soil frozen depth on vegetation growth in the source region of the Yellow River for 1982–2015. *Theoretical and Applied Climatology, 140*, 1185–1197.

Wen, J., Zhang, J. Q., Nie, Z. L., Zhong, Y., & Sun, H. (2014). Evolutionary diversifications of plants on the Qinghai-Tibetan Plateau. *Frontiers in Genetics, 5*, 4.

Xiao, J. (2014). Satellite evidence for significant biophysical consequences of the "grain for green" program on the loess plateau in China. *Journal of Geophysical Research: Biogeosciences, 119*(12), 2261–2275.

Xie, S. C., Wang, Z. Y., Wang, H. M., Chen, F. H., & An, C. B. (2002). The occurrence of a grassy vegetation over the Chinese loess plateau since the last interglacier: The molecular fossil record. *Science in China (D), 45*, 53–62.

Yan, Q., Zhu, J., Zheng, X., & Jin, C. (2015). Causal effects of shelter forests and water factors on desertification control during 2000–2010 at the Horqin Sandy land region, China. *Journal of Forestry Research, 26*(1), 33–45.

Yang, M. X., Nelson, F. E., Shiklomanov, N. I., Guo, D. L., & Wan, G. N. (2010). Permafrost degradation and its environmental effects on the Tibetan plateau: A review of recent research. *Earth Science Reviews, 103*, 31–44.

Yi, S., Saito, Y., Oshima, H., Zhou, Y., & Wei, H. (2003). Holocene environmental history inferred from pollen assemblages in the Huanghe (Yellow River) delta, China: Climatic change and human impact. *Quaternary Science Reviews, 22*(5–7), 609–628.

Yu, J., Fu, Y., Li, Y., Han, G., Wang, Y., Zhou, D., Sun, W., Gao, Y., & Meixner, F. X. (2011). Effects of water discharge and sediment load on evolution of modern Yellow River Delta, China, over the period from 1976 to 2009. *Biogeosciences, 8*(9), 2427–2435.

Yu, Y., Yu, P., Wang, Y., Wan, Y., Wang, B., Han, X., Tu, X., Li, J., Xu, L., Wang, X., & Liu, Z. (2023). Natural revegetation has dominated annual runoff reduction since the grain for green program began in the Jing River basin, Northwest China. *Journal of Hydrology, 625*, 129978.

Zeng, Y. N., Feng, Z. D., & Cao, G. C. (2003). Land cover change and its environmental impact in the upper reaches of the Yellow River, Northeast Qinghai-Tibetan plateau. *Mountain Research and Revelopment, 23*(4), 353–361.

Zhai, J., Wang, L., Liu, Y., Wang, C., & Mao, X. (2023). Assessing the effects of China's three-north shelter forest program over 40 years. *Science of the Total Environment, 857*, 159354.

Zhang, K. (1989). *Tectonics and resources of Ordos fault-block* (p. 394). Shanxi Science and Technology Press. (in Chinese).
Zhang, L., Yue, L., & Xia, B. (2003). The study of land desertification in transitional zones between the MU US Desert and the loess plateau using RS and GIS—A case study of the Yulin region. *Environmental Geology, 44*, 530–534.
Zhang, Q. Q., Ferguson, D. K., Mosbrugger, V., Wang, Y. F., & Li, C. S. (2012). Vegetation and climatic changes of SW China in response to the uplift of Tibetan plateau. *Palaeogeography, Palaeoclimatology, Palaeoecology, 363*, 23–36.
Zhang, D., Zuo, X., & Zang, C. (2021). Assessment of future potential carbon sequestration and water consumption in the construction area of the three-north shelterbelt Programme in China. *Agricultural and Forest Meteorology, 303*, 108377.
Zhang, B., Tian, L., Yang, Y., & He, X. (2022). Revegetation does not decrease water yield in the loess plateau of China. *Geophysical Research Letters, 49*(9), e2022GL098025.
Zhao, Y., Chen, F., Zhou, A., Yu, Z., & Zhang, K. (2010). Vegetation history, climate change and human activities over the last 6200 years on the Liupan Mountains in the southwestern loess plateau in Central China. *Palaeogeography, Palaeoclimatology, Palaeoecology, 293*(1–2), 197–205.
Zheng, F. L. (2006). Effect of vegetation changes on soil erosion on the loess plateau. *Pedosphere, 16*(4), 420–427.
Zhou, H., Zhao, X., Tang, Y., Gu, S., & Zhou, L. (2005). Alpine grassland degradation and its control in the source region of the Yangtze and Yellow Rivers, China. *Grassland Science, 51*(3), 191–203.
Zhou, D., Zhao, S., & Zhu, C. (2012). The grain for green project induced land cover change in the loess plateau: A case study with Ansai County, Shanxi Province, China. *Ecological Indicators, 23*, 88–94.
Zhu, Z. Y. (1989). The formation of river terraces and evolution of drainage system in the middle Yellow River. *Acta Geographica Sinica, 44*(4), 429–440.

Chapter 8
Flood and Prevention

Contents

8.1 Introduction

The Yellow River has earned "China's sorrow" for its frequent flood events in the history (Qian and Zhou, 1965). According to the record of the historical documents, there have been over 1000 flood events in the past 4000 years (Chen et al., 2012). The floods caused the shift of river channels, overspill, breach of levees in the lower reaches. The inundation of croplands in the surrounding plains led to the significant loss of life, destruction of properties, crop failures, and social instability in history. For examples, the extreme flood in AD 1642 destroyed the inner city of the former capital, Kaifeng, the dynastic China, killing around 300,000 people (Storozum et al., 2020). The 1931 floods caused a death toll between 900,000 to 2,000,000 people, and the 1938 floods killed 3–4 million people (Duan et al., 2016). Floods also caused frequent shifts of the lower channels. During the mid seventeenth century the breach frequency was as high as three times per year on average. Over the past millennium, the river's lower course shifted approximately every 25 years, with levee breaches occurring on average once per year (Chen et al., 2012). Floods in the middle reaches of the Yellow River basin, caused by extreme rainstorms, can

T. Zhu, *Watershed Environmental Degradation and Sustainable Development*,
https://doi.org/10.1007/978-3-032-02448-0_8

damage terrace lands and check dam areas. For instance, on July 26, 2017, a flood triggered by a storm with a 500-year recurrence period struck the northern Loess Plateau. This event caused severe water erosion, mass movement, and significant damage to old terraces, small earthen check dams, and roads. Sediment-laden water flows into the Yellow River, with some of the sediment settling in the lower river channel. This accumulation raises the riverbed above the surrounding North China Plain, creating a "perched river" and further heightening the risk of flooding. This chapter will explore the causes of floods and strategies for their mitigation and prevention.

8.2 Flood Frequency and Its Causes

A variety of methods are used to determine flood frequency and magnitude in the Yellow River basin. The Yellow River Hydrologic Bureau operates 32 hydrological stations that monitor flow discharge, each with a record spanning at least 50 years (Bai et al., 2016). In addition, historical documents provide accounts of major flood events over the past 2,000–3000 years, including those caused by levee breaches and channel avulsions (Li et al., 2021). Paleo-flood events have also been reconstructed through the analysis of slack-water deposits preserved within loess-paleosol sequences and archaeological cultural layers (Yu et al., 2023).

Flood frequency in the Yellow River basin has varied over time. During the prehistoric period, it was primarily influenced by climatic trends; however, with population growth and increasing human activity, anthropogenic factors have become the dominant drivers of flood frequency. Yu et al. (2023) analyzed a 12,000-year record of Yellow River flood events by integrating sedimentary evidence with historical documentation of levee overtoppings and breaches. Their study identified four distinctive phases: (1) 10,000 to 5000 BCE-the climate dominated; (2) 5000 to 2500 BCE - transition from a climate-dominated to a human-dominated system;(3) 2500 to 500 BCE -rapidly enhancing human disturbances; (4) 500 BCE to 2000 CE - dominance of anthropogenic influences over climate on the system. Their findings indicate that flood events became nearly an order of magnitude more frequent over the past millennium compared to the mid-Holocene, largely due to intensified human activities in the basin—particularly in the Loess Plateau. Similarly, studies have shown that agricultural land use in the Upper Mississippi River Valley has led to significant increases in flood magnitudes—rising by one to two orders of magnitude compared to pre-agricultural, natural conditions (Knox, 2001).

The Yellow River basin is characterized by widely varying levels of human disturbance and significant temporal and spatial variability in precipitation. Although major floods have often occurred in the lower reaches of the Yellow River basin, the flood water is mainly from the upper and middle basin. Due to the relatively low intensity of human activity in the source region, climate trends remain the primary factor influencing flow discharge. Flow generation in the middle reaches, the Loess Plateau, attributed to the distinctive physical and human-induced conditions. The

Loess Plateau is in temperate monsoon climates, ranging from semi-arid to semi-humid climates. Precipitation is marked by significant seasonal variation, with more than 70% of the annual total occurring during the rainy season, from May to September (Fig. 8.1). These summer storms, heavily influenced by monsoonal shifts, are typically intense (Figs. 8.2, 8.3 and 8.4).

The large year-to-year variation in precipitation contributes significantly to flood occurrences. According to hydrologic data collected by the Yellow River Commission (Yu, 2024), Figure 8.5 presents the maximum and mean monthly precipitation from 1952 to 2019 across 16 locations (Fig. 8.4) along the Yellow River, spanning from its source to its mouth. The mean ratio of maximum to mean monthly precipitation is 7.28, with values ranging from 4.27 to 10.25. Figure 8.6 displays the mean, maximum, and minimum annual precipitation over the same period, showing a ratio of maximum to minimum yearly precipitation of 3.2, with a range from 1.64 to 5.12.

There have been many studies on the temporal changes of annual precipitation and rainfall patterns in the basin. For examples, Fu et al. (2004) reported that annual precipitation has remained stable over the last 50 years. Wang et al. (2013) found no significant increase in extreme precipitation events within the basin. Zhao et al. (2018) reported that more weather stations on the Loess Plateau show a decreasing trend in extreme daily precipitation than those indicating an increase.

Infiltration-excess flow process is predominant in the region. Thus, rainfall intensity and infiltration rates determine the flow rates. The Loess Plateau region is highly susceptible to surface flow generation due to intense summer storms and low infiltration rates in cropland and gully areas. There have been numerous studies on the infiltration rates and the factors affecting them, such as landuse, slope gradients, soils, and land management practices (Li et al., 1995; Fang et al., 2008; Zhao et al.,

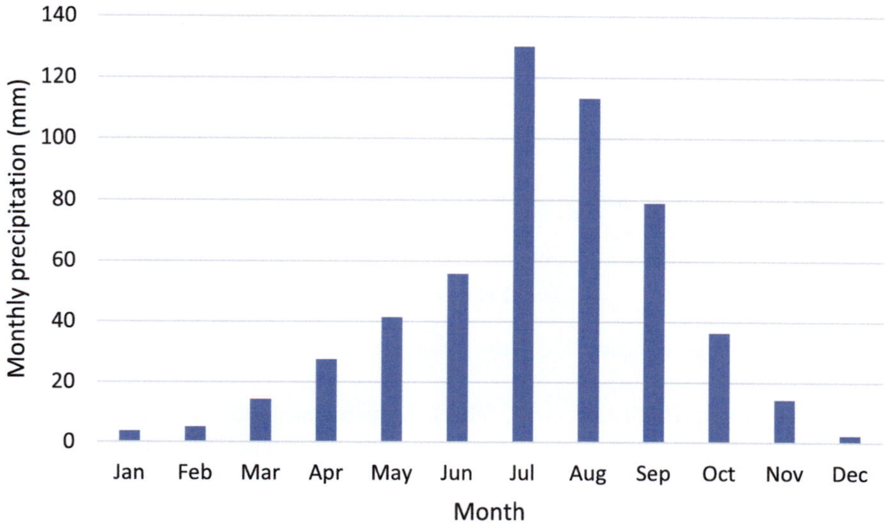

Fig. 8.1 Mean monthly precipitation in the WJG watershed on the Loess Plateau

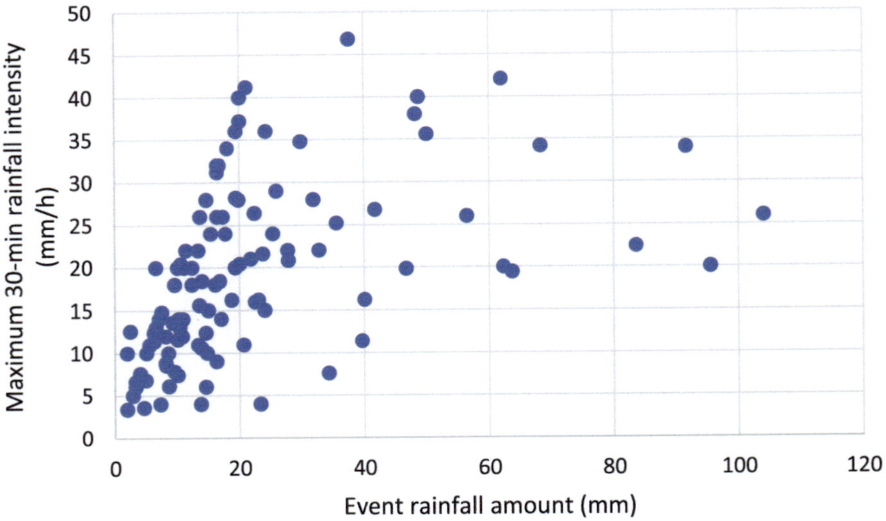

Fig. 8.2 Maximum 30-min rainfall intensity for runoff-generating storms in the WJG watershed on the Loess Plateau

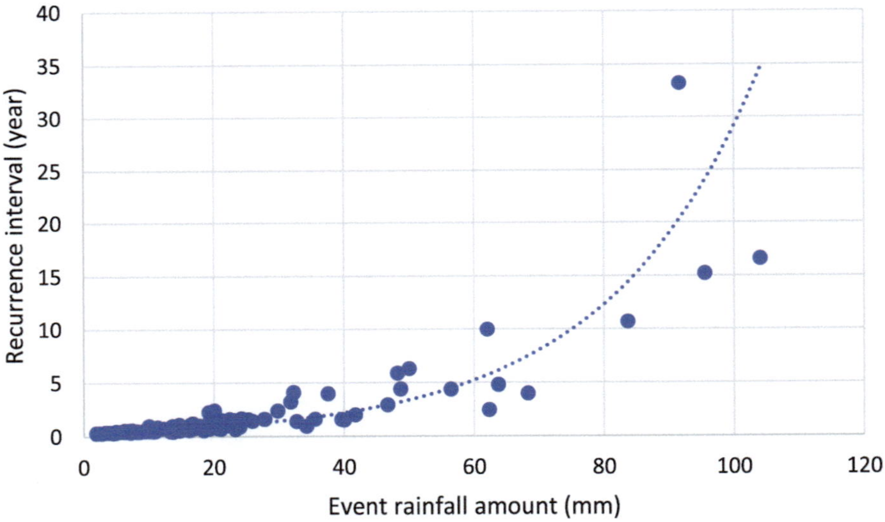

Fig. 8.3 Event rainfall amount with different recurrence intervals

2013; Wang et al., 2015; Dan et al., 2023). Terrace lands and vegetated areas typically exhibit significantly higher infiltration rates compared to sloped cropland, resulting in reduced overland flow. According to Zhu and Zhu (2014), terrace land and grassland produced 30.3% and 15.2%, respectively, of the average annual runoff observed on cropland.

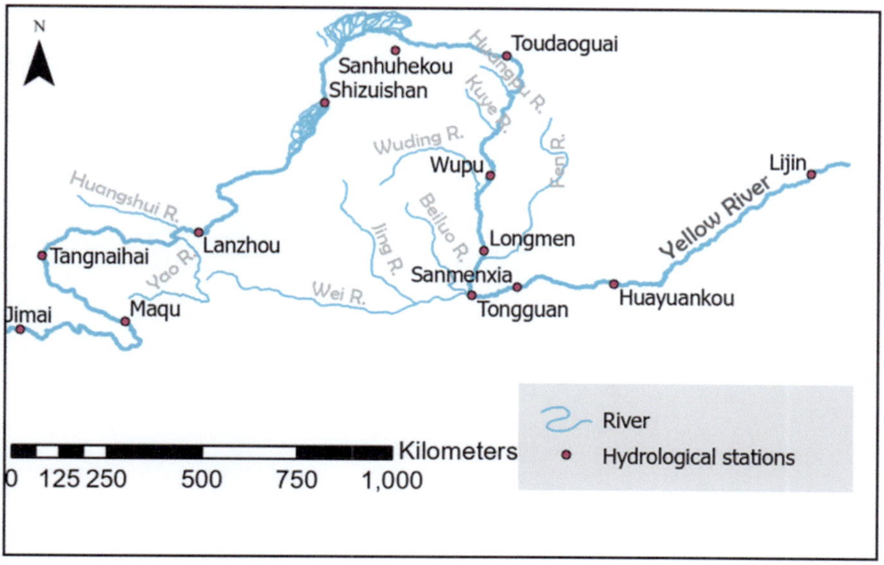

Fig. 8.4 Map of the locations

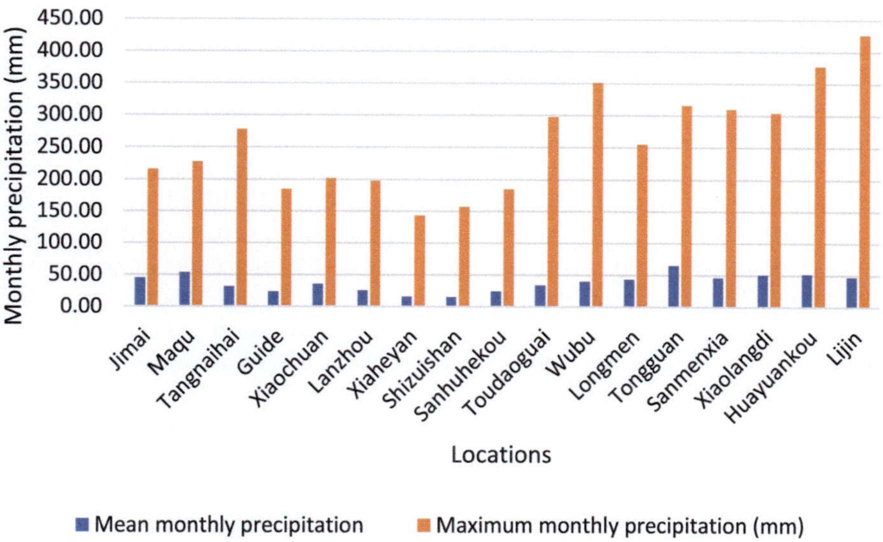

Fig. 8.5 Mean monthly precipitation and maximum monthly precipitation in the selected locations of the Yellow River basin

Topographically, the dissected loess plateau accelerates water flow into the river, while the steep longitudinal profiles of the riverbed enable rapid movement of floodwaters through the middle basin and the Jinshaan Gorge. This gorge, approximately 700 km long, cuts through the eastern Ordos region, flowing north to south. Upon

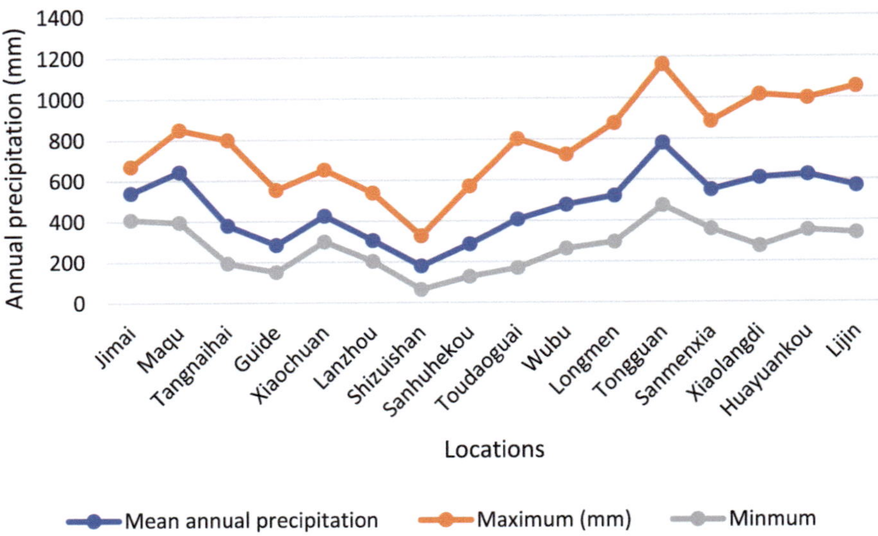

Fig. 8.6 Mean yearly precipitation and maximum and minimum yearly precipitation in the selected locations of the Yellow River basin

transitioning from the mountainous areas to the plains, the riverbed slope decreases significantly, averaging just 0.012% in the lower reaches of the North China Plain. The lower channel of the Yellow River is elevated due to the sedimentation in the channel. The 800-km long confined lower channel by levees has limited capacity for flooding water to pass through due to shallow depths and lower longitudinal gradients. The channel depth under the bankfull condition is only about 2 m. The distance between the levee ranged from 1.4 km to 8.5 km (Qian et al., 1987; Wu et al., 2005). This creates extreme flood vulnerability. Consequently, levees were frequently breached in recent history. Since the founding of the Peoples Republic of China in 1949, one of the primary efforts to prevent devastating floods has been to raise and reinforce the 1,300 km of levee embankments along the north and south banks of the lower channel. However, this approach is costly and may increase the risk of future flood hazards.

While floods are triggered by excessive water flow, sediment deposition in the lower river channel is a fundamental underlying cause. The sediments accumulating in the riverbed of the lower reaches originate almost entirely from the middle basin, the Loess Plateau region. Loess is believed to origin from central Asia by wind-blown deposits in the Quaternary period (Liu, 1964). Accordingly, the closer to the source region, the coarser the loess is. Three loess belts can be divided from north-west to southeast of the Loess Plateau: coarse, silt and clay loess belt. The long-term sediment load measured at Sanmenxia station (1919–1960) averaged 1.6 billion tons per year, among which 0.4 billion tons were deposited in the lower channel. The size of deposited sediment has a higher percentage of coarse sediments with a diameter greater than 0.05 mm (Xu & Cheng, 2002; Xu et al., 2006); The source area of coarse sediment in the Yellow River is predominately coming from the coarse loess belt. Over 70% of the total coarse sediment is from the area of 110,000 km^2, among which 50,000 km^2 contributed more than half of the coarse sediment. In the coarse loess belt, the wind and surface water erosion make the surface loess gradually coarsening due to fine particles are selectively transported by wind and surface runoff.

The transport of coarse sediments also exhibits seasonal variability. Xu (2000) observed that the proportion of coarse sediments in tributary rivers exhibits a double-peaked pattern: one peak occurs during the low-water season, and the other corresponds to hyperconcentrated flows during the high-water season. Chien et al. (1979) defined hyperconcentrated flows as having sediment concentrations greater than 400 kg/m^3. The formation of these flows is believed to depend on factors such as the content of fine particles (clay and silt), flow shear stress, and vertical velocity gradients (Chien & Ma, 1958; Chien & Wan, 1983; Xu, 1998, 2000). Overall, the coarse sediment load is significantly higher during the high-water season than in the low-water season. This indicates that rainfall during the high-water season (summer) plays a crucial role in determining the amount of coarse sediment delivered to the Yellow River. This temporal variation is further explained by historical patterns: during the 1960s, when summer rainfall was abundant, higher amounts of coarse sediment were transported to the Yellow River. However, coarse sediment loads have declined since the 1970s, which is ascribed to a decrease in summer precipitation (Xu, 2000).

8.3 Flood Mitigation and Prevention

Flood mitigation and prevention in the lower channel of the Yellow River has been among the most important and difficult tasks in China. Efforts have been made since at least 3000 years ago. The major flood control measures in ancient times involved in engineering approaches, mainly constructing levee embarkments to confine flood water between levees. The construction of the levees along the lower channel can be traced back to at least the Spring and Autumn Period (770–476 BC). Various strategies had been attempted to control the sediment-laden floods during the historical periods, such as clearing sediments with converging flow, storing clear and release turbid water, as well as protecting levees through the deposition of the sediment bars. Nevertheless, there had been frequent levee breaches before the founding of the PRC in 1949. Over the period of 602 B.C. to 1949, the channel had 1593 levee breaches, and the major course of the river channel shifted 26 times by avulsion (Wang & Liu, 2019), though a recent study showed only 588 levee breaches over the period of AD 11 to 1949 (Li et al., 2021). The last major levee breaches occurred in 1938 when the Chinese Nationalist Party Army deliberately breached the levee at Huayuankou to prevent the Japanese invasion. The Huayuankou breach was closed in 1947, and the levee was reconstructed to along its post-1855 channel course (Shu & Finlayson, 1993). The 1855 flood changed the course of Yellow River drastically, redirecting its discharge from the Yellow Sea to the Bohai Gulf, shifting about 300 miles northward.

Since the 1950s, a comprehensive flood control strategy has been implemented, combining engineering, biological, and agronomic measures, while coordinating flood control efforts from a basin-wide perspective, as discussed below.

8.3.1 Soil and Water Conservation in the Middle Basin

While the flood hazards mainly occur in the lower channel, the root cause of the floods is in the middle basin that is the main source region of flood water and coarse sediments. Thus, reduction of water and sediment entering the river channel in the middle reaches is the key to preventing flood hazards in the lower basin.

To reduce runoff generation, land use changes and agronomical measures have been implemented at the small watershed scale, including converting steep slopeland to terrace land, increasing forest land and grass land. Currently, there is about 9.02 million hectares of terrace lands constructed in the Loess Plateau region (Lu et al., 2023). Large scales of revegetation projects have been implemented by the Chinese government, especially since 1999 when the "Grain for Green" project started. Remote sensing data showed that the vegetation cover in 12 subwatersheds of the Loess Plateau increased from 25% in 1978, to 29% in 1998, and 40% in 2010 (Wang et al., 2016). It is important to assess how effective those soil conservation measures in reducing runoff under extreme rainfall events. Zhu and Zhu (2014) reported that during extreme storms with a return period of 10 years or more, the average event runoff from terraced land is only 25.4% of the runoff produced from sloped cropland. However, the runoff from forestland is 75.8% of the runoff generated on sloped cropland, making it significantly less effective at reducing runoff compared to terraced land.

Since 1950s, nine large reservoirs with storage capacities exceeding 1 km^3 have been constructed along the Yellow River. Among them, four reservoirs—Liujiaxia, Longyangxia, Sanmenxia, and Xiaolangdi—each with a capacity greater than 5 km^3, are particularly effective in significantly reducing flood peaks (Bai et al., 2016; Wang et al., 2006). In addition, numerous check dams have been built in the Loess Plateau region. According to the Yellow River Commission, there were 6057 large dams, 11881 medium dams, and 74,820 small dams were built by 2011 in the region (Wang et al., 2021). Those check dams can effectively retain and delay runoff during flood periods (Xu et al., 2004; Guo et al., 2021; Yuan et al., 2022). For example, the check dam systems constructed in the Huangpuchuan River and Wuding River, two main tributaries in the middle reach of the Yellow River, are estimated to reduce runoff by 39% and 12%, respectively (Shi et al., 2016; Shi et al., 2019).

The spatial arrangement of various conservation measures within a small watershed plays a crucial role in reducing flow discharges. Typically, terraces are constructed on upper hillslopes with gentle to moderate gradients, while shrubs, grasses, and trees are planted on the steeper lower gully slopes. Check dams are strategically built at the base of gullies to further manage water flow. Zhu (2016) analyzed flow discharges in the twin catchments, one with conservation measures and one without, during rainstorm events of varying return periods. In the managed catchment, event flow discharges were reduced by 53.8%, 64.1%, 34.7%, 64.9%, 47.6%, and 50.7% for return periods of <1, 1–2, 2–5, 5–10, 10–20, and > 20 years, respectively. Similarly, the peak flow discharge reduction was 69.3%, 70.6%, 50.6%, 72.8%, 58.4%, and 15.8%, demonstrating the effectiveness of the conservation measures.

8.3.2 Water and Sediment Management by Reservoirs

Reservoirs were constructed across the Yellow River to store flood water, such as Sanmenxia and Xiaolangdi reservoirs. The Sanmenxia Dam, located in Henan Province, was constructed over the period of 1957–1960. The dam is a concrete gravity dam with a heigh of 353 m above sea level. The drainage area above the dam accounts for 92% of the Yellow River and controls 89% of the total flow discharge and 98% of the total sediment load in the Yellow River (Wang et al., 2005; Wang et al., 2007). Due to the severe sedimentation behind the dam, the flood water storage capacity rapidly decreased after its impoundment, with 1.8 billion tons of sediment trapped after only 18 months of its operation (Morris & Fan, 1998). As the reservoir is located in the confluence of the Wei river, the largest tributary of the Yellow River, and the main channel of the Yellow River, the backwater sediment deposition along the Wei river, has posed more severe flood threat to the lower stream of the Wei river where Xi'an city, the ancient Chinese capital city with a population of over than 8 million people. Due to the severe sedimentation, the dam was reengineered by reconstructing the dam's outlet for better sediment management and flood control. The reservoir operating mode changed from the early mode of retaining flood water in the rainy season to the mode of retaining clear water in the dry season and discharging turbid water in the flood season, which, to certain degree, defy the original purpose of the reservoir (Pan, 1991; Wang et al., 2005).

Learned lesson from the Sanmenxia reservoir, the Chinese government built another dam, Xiaolangdi, about 130 km downstream of the Sammenxia reservoir, over the period of 1991 to 2001. It drains 92.3% of the total Yellow River basin, and controls 100% of sediment. The crest of the dam is 281 m above sea level, with a total storage capacity of 12.65×10^9 m^3 (Kong et al., 2017). The construction of the Xiaolongdi reservoir aims at further regulating water and sediment, as well as hydro-power generation. To manage the flood water, the reservoir can reduce the peak flow from 335000 m^3/sec to 220000 m^3/sec during the flood season. To manage the sediment, the clear water with low sediment concentration in the reservoir during the low water season is released as high-speed jet flows to scour the riverbed, whereas the bottom turbid water with high sediment concentration during the high-water season is discharged from the reservoir to maintain the storage capacity of the reservoir (Hu et al., 2012; Xia et al., 2016). Through managing water and sediment, the riverbed downstream has been lowered by more than 1 m, which reversed the trend of mean annual sedimentation rate of 10 cm prior to the operation. By 2020, the actual sedimentation rate in the reservoir was about 42% of the anticipated sedimentation one by design.

8.3.3 *Levees Along the Lower Channel*

The levees built along the lower channel aimed to confine river water between the levees. The lower channel was classified into upper, middle and lower sections (Qian et al., 1993). The upper section is a braided channel with a length of 299 km, a width of main channel ranging from 1–4.5 km, and mean depth of 2 m under bank-full conditions. The middle section is 165 km and is a transition from braided channel to meandering channel, with a width of main channel from 0.5–1.6 km. The distance between levees ranging from 1.4 to 8.5 km, a mean river bed slope of 0.00015; The lower section is mainly meandering channel with a length of 300 km, a mean river slope of 0.0001, and the distance between levees is 0.45–5 km (Wu et al., 2005). The total length lining the south and north bank of the lower channel is more than 1300 km.

Since the founding of the PRC in 1949, there has not been a levee breach due to the comprehensive levee restoration projects. There have been four major levee restoration periods after 1949: 1950–1957, 1962–1965, 1974–1985, and 1996–2018 (Hu, 2020). During each of the period, levees were raised approximately 1 m along their whole length. Various engineering works have been implemented to protect the safety of levees and regulate the river flow paths, including levee protection works constructed at the vulnerable sections of the levees and the concave banks of river bands, which prevent the levees from flood water scouring.

8.3.4 *Flood Retention*

There are two major flood retention basins in the upper section of the lower river channels, one is the Lake Dongping retention basin with a total area of 627 km^2 (Li & Cao, 1990) and the present flood storage capacity is 3.02×10^9 m^3. The other is the North Golden Levee basin with a total area of 2316 km^2 but has not been used for flood water storage due to the large population and damage it can cause (Shu & Finlayson, 1993). During an extreme flood event, both can be used to divert flood water from the main channel after the residents are evacuated from the basins.

8.4 Conclusion

Although the frequent and devastating floods in the lower Yellow River throughout history were caused by water, the underlying root cause was sediment. The sedimentation in the lower channel created a "perched river" that is shallow and unstable. Since more than 90% of the total sediment load in the Yellow River is from the middle reaches, the Loess Plateau region, managing floodwater in the lower channel must be coordinated with controlling soil erosion in the middle reaches. Soil and water conservation efforts in the Loess Plateau have been crucial in reducing the sediment and runoff flowing into the lower channel. The dramatic reduction of sediments, combined with sediment flushing techniques, has made siltation of the reservoirs less severe, allowing the reservoirs to retain water during flood periods. The levee constructed along the lower channel, along with reinforcement, protects the North China Plain from flood hazards. In extreme flood events, while systematic regulation of the reservoirs is essential, flood retention basins in the lower reaches may be used for excessive water storage.

References

Bai, P., Liu, X., Liang, K., & Liu, C. (2016). Investigation of changes in the annual maximum flood in the Yellow River basin, China. *Quaternary International, 392*, 168–177.

Chen, Y., Syvitski, J. P., Gao, S., Overeem, I., & Kettner, A. J. (2012). Socio-economic impacts on flooding: A 4000-year history of the Yellow River, China. *Ambio, 41*, 682–698.

Chien, N., & Ma, H. (1958). *The viscosity of stress-flow relation of sediment-laden flow Sediment Research, 3*(3), 52–77. (in Chinese).

Chien, N., & Wan, Z. (1983). *Sediment movement mechanics*. Science Press.

Chien, N., Wan, Z., & Qian, Y. (1979). Hyperconcentrated flows in the Yellow River and its tributaries. *Journal of Tsinghua University, 19*(2), 1–17.

Dan, C., Liu, G., Zhao, Y., Shu, C., Shen, E., Liu, C., Tan, Q., Zhang, Q., Guo, Z., & Zhang, Y. (2023). The effects of typical grass cover combined with biocrusts on slope hydrology and soil erosion during rainstorms on the loess plateau of China: An experimental study. *Hydrological Processes, 37*(1), e14794.

Duan, W., He, B., Nover, D., Fan, J., Yang, G., Chen, W., Meng, H., & Liu, C. (2016). Floods and associated socioeconomic damages in China over the last century. *Natural Hazards, 82*, 401–413.

Fang, H. Y., Cai, Q. G., Chen, H., & Li, Q. Y. (2008). Effect of rainfall regime and slope on runoff in a gullied loess region on the loess plateau in China. *Environmental Management, 42*, 402–411.

Fu, G., Chen, S., Liu, C., & Shepard, D. (2004). Hydro-climatic trends of the Yellow River Basin for the last 50 years. *Climate Change, 65*(1–2), 149–178.

Guo, W. Z., Wang, W. L., Xu, Q., Hu, J. J., & Zhu, L. L. (2021). Distribution, failure risk and reinforcement necessity of check-dams on the loess plateau: A review. *Journal of Mountain Science, 18*(2), 499–509.

Hu, Y. S. (2020). Review of levee restorations in the downstream of Yellow River throughout 70 years. *Yellow River, 42*(6), 18–21. (In Chinese).

Hu, P., Cao, Z., Pender, G., & Tan, G. (2012). Numerical modelling of turbidity currents in the Xiaolangdi reservoir, Yellow River, China. *Journal of Hydrology, 464*, 41–53.

Knox, J. C. (2001). Agricultural influence on landscape sensitivity in the upper Mississippi River valley. *Catena, 42*(2–4), 193–224. https://doi.org/10.1016/S0341-8162(00)00138-7

Kong, D., Miao, C., Wu, J., Borthwick, A. G., Duan, Q., & Zhang, X. (2017). Environmental impact assessments of the Xiaolangdi reservoir on the most hyperconcentrated laden river, Yellow River, China. *Environmental Science and Pollution Research, 24*, 4337–4351.

Li, G. Y., & Cao, J. F. (1990). Economic analysis of flood control on the middle and lower Yellow River in the post-1949 period. *Yellow River, 12*(1), 13–18. (in Chinese).

Li, G., Luk, S. H., & Cai, Q. G. (1995). Topographic zonation of infiltration in the hilly loess region, North China. *Hydrological Processes, 9*(2), 227–235.

Li, W. J., Yu, S. Y., Pan, J., Cao, X., Chen, Y., & Wang, Y. (2021). A 2000-year documentary record of levee breaches on the lower Yellow River and their relationship with climate changes and human activities. *The Holocene, 31*(3), 333–345.

Liu, D. S. (1964). *Loess in the middle Yellow River Drainage Basin*. Science Press.

Lu, Y., Li, X., Xin, L., Song, H., & Wang, X. (2023). Mapping the terraces on the loess plateau based on a deep learning-based model at 1.89 m resolution. *Scientific Data, 10*(1), 115.

Morris, G. L., & Fan, J. (1998). *Reservoir sedimentation handbook: Design and management of dams, reservoirs, and watersheds for sustainable use*. McGraw Hill.

Pan, X. D. (1991). The main laws of scouring and deposition of the lower river course and the reservoir area of Sanmenxia. *Yellow River, 13*(11), 13–20. (In Chinese).

Qian (Chien), N, & Zhou, W. H. (1965). *Fluvial processes in the lower Yellow River*. Science Press.

Qian (Chien), N, Zhang, R., & Zhou, Z. D. (1987). *Fluvial processes*. Science Press.

Qian, Y. Y., Ye, Q. C., & Zhou, Z. D. (1993). *Changes in discharge and sediment load in the main stem of the Yellow River and associated fluvial processes*. China Building Material Industrial Press. (in Chinese).

Shi, H., Li, T., Wang, K., Zhang, A., Wang, G., & Fu, X. (2016). Physically based simulation of the streamflow decrease caused by sediment-trapping dams in the middle Yellow River. *Hydrological Processes, 30*(5), 783–794.

Shi, P., Zhang, Y., Ren, Z., Yu, Y., Li, P., & Gong, J. (2019). Land-use changes and check dams reducing runoff and sediment yield on the loess plateau of China. *Science of the Total Environment, 664*, 984–994.

Shu, L., & Finlayson, B. (1993). Flood management on the lower Yellow River: Hydrological and geomorphological perspectives. *Sedimentary Geology, 85*(1–4), 285–296.

Storozum, M., Lu, P., Wang, S., et al. (2020). Geoarchaeological evidence of the AD 1642 Yellow River floodthat destroyed Kaifeng, a former capital of dynastic China. *Scientific Reports, 10*, 3765.

Wang, Z., & Liu, C. (2019). Two-thousand years of debates and practices of Yellow River training strategies. *International Journal of Sediment Research, 34*(1), 73–83.

Wang, G., Wu, B., & Wang, Z. Y. (2005). Sedimentation problems and management strategies of Sanmenxia reservoir, yellow river, China. *Water Resources Research, 41*(9), 1.

Wang, H., Yang, Z., Saito, Y., Liu, J. P., & Sun, X. (2006). Interannual and seasonal variation of the Huanghe (Yellow River) water discharge over the past 50 years: Connections to impacts from ENSO events and dams. *Global and Planetary Change, 50*(3–4), 212–225.

Wang, Z. Y., Wu, B., & Wang, G. (2007). Fluvial processes and morphological response in the yellow and Weihe Rivers to closure and operation of Sanmenxia dam. *Geomorphology, 91*(1–2), 65–79.

Wang, W., Shao, Q., Yang, T., Peng, S., Yu, Z., Taylor, J., Xing, W., Zhao, C., & Sun, F. (2013). Changes in daily temperature and precipitation extremes in the Yellow River Basin, China. *Stochastic Environmental Research and Risk Assessment, 27*, 401–421.

Wang, L., Zhong, C., Gao, P., Xi, W., & Zhang, S. (2015). Soil infiltration characteristics in agro-forestry systems and their relationships with the temporal distribution of rainfall on the loess plateau in China. *PLoS One, 10*(4), e0124767.

Wang, S., Fu, B., Piao, S., Lü, Y., Ciais, P., Feng, X., & Wang, Y. (2016). Reduced sediment transport in the Yellow River due to anthropogenic changes. *Nature Geoscience, 9*(1), 38–41.

Wang, Z., Chen, Z., Yu, S., Zhang, Q., Wang, Y., & Hao, J. (2021). Erosion-control mechanism of sediment check dams on the loess plateau. *International Journal of Sediment Research, 36*(5), 668–677.

Wu, B., Wang, G., Ma, J., & Zhang, R. (2005). Case study: River training and its effects on fluvial processes in the lower Yellow River, China. *Journal of Hydraulic Engineering, 131*(2), 85–96.

Xia, X., Dong, J., Wang, M., Xie, H., Xia, N., Li, H., Zhang, X., Mou, X., Wen, J., & Bao, Y. (2016). Effect of water-sediment regulation of the Xiaolangdi reservoir on the concentrations, characteristics, and fluxes of suspended sediment and organic carbon in the Yellow River. *Science of the Total Environment, 571*, 487–497.

Xu, J. (1998). A study of physico-geographical factors for formation of hyperconcentrated flows in the loess plateau of China. *Geomorphology, 24*(2–3), 245–255.

Xu, J. (2000). Grain-size characteristics of suspended sediment in the Yellow River, China. *Catena, 38*(3), 243–263.

Xu, J., & Cheng, D. (2002). Relation between the erosion and sedimentation zones in the Yellow River, China. *Geomorphology, 48*(4), 365–382.

Xu, X. Z., Zhang, H. W., & Ouyang, Z. (2004). Development of check-dam systems in gullies on the loess plateau, China. *Environmental Science & Policy, 7*(2), 79–86.

Xu, J., Yang, J., & Yan, Y. (2006). Erosion and sediment yields as influenced by coupled eolian and fluvial processes: The Yellow River, China. *Geomorphology, 73*(1–2), 1–15.

Yu, Y. (2024). Water and sediment data from the Yellow River, China. *Mendeley Data, V1*, 1. https://doi.org/10.17632/grzcrfjvjr.1

Yu, S. Y., Li, W. J., Zhou, L., Yu, X., Zhang, Q., & Shen, Z. (2023). Human disturbances dominated the unprecedentedly high frequency of Yellow River flood over the last millennium. *Science Advances, 9*(8), eadf8576.

Yuan, S., Li, Z., Chen, L., Li, P., & Zhang, Z. (2022). Influence of check dams on flood hydrology across varying stages of their lifespan in a highly erodible catchment, Loess Plateau of China. *Catena, 210*, 105864.

Zhao, L., Wang, L., Liang, X., Wang, J., & Wu, F. (2013). Soil surface roughness effects on infiltration process of a cultivated slopes on the loess plateau of China. *Water Resources Management, 27*, 4759–4771.

Zhao, G., Zhai, J., Tian, P., Zhang, L., Mu, X., An, Z., & Han, M. (2018). Variations in extreme precipitation on the loess plateau using a high-resolution dataset and their linkages with atmospheric circulation indices. *Theoretical and Applied Climatology, 133*, 1235–1247.

Zhu, T. X. (2016). Effectiveness of Conservation Measures in Reducing Runoff and Soil Loss Under Different Magnitude–Frequency Storms at Plot and Catchment Scales in the Semi-arid Agricultural Landscape. *Environmental Management, 57*, 671–682.

Zhu, T. X., & Zhu, A. X. (2014). Assessment of soil erosion and conservation on agricultural sloping lands using plot data in the semi-arid hilly loess region of China. *Journal of Hydrology: Regional Studies, 2*, 69–83.

Chapter 9
Droughts, Salinization and Sustainable Agriculture

Contents

9.1 Introduction

Drought refers to deficiency of precipitation over an extended period of time, and all life must adjust to the unexpected shortage of water. Drought can be classified into meteorological drought, agricultural drought, hydrological drought, and socioeconomic drought. While droughts can occur in any climate, they are more common in semi-arid climates than in other climates. The majority of the Yellow River basin has a semi-arid climate, marked by significant temporal variability in rainfall. With a population of 110 million, agriculture plays a crucial role in sustaining the people living in the basin. In the middle reaches of the Yellow River basin, the Loess Plateau, most agricultural land relies on rain-fed irrigation. Droughts significantly impact food production by reducing soil moisture availability during the critical

T. Zhu, *Watershed Environmental Degradation and Sustainable Development*,
https://doi.org/10.1007/978-3-032-02448-0_9

crop growth season. Climate changes could also potentially affect the frequency and severity of droughts. In this chapter, we will first describe the some of the major drought indexes. Second, we will discuss the spatial and temporal variations of droughts in the Yellow River basin and their impact on environment and agriculture. Third, we will recommend the strategies and methods for drought mitigation. Finally, we will discuss the salization that is sometimes related to the drought mitigation such as irrigation in the region, and its impact and remediation methods.

9.2 Drought Indexes

To evaluate drought frequency and severity, various drought indices have been applied to the Yellow River basin, such as percentile of normal, the Standard Precipitation Index (SPI), the Palmer Drought Index (PDI), and the Standardized Precipitation Evapotranspiration Index (SPEI). Each index comes with its own set of advantages and limitations, which will be discussed below.

9.2.1 Percentile of Normal

This is to calculate precipitation over a certain period of time as a percentage of the mean precipitation on a long-term basis (i.e. 30 years). This index is simple and frequently used to assess the temporal variation of precipitation in the Yellow River basin at various time scales (Wang et al., 2012a, 2012b; Zhao et al., 2015; Sun et al., 2015; Tang et al., 2018). It is effective for comparing conditions within a single region or season but poses challenges when used to compare different regions.

9.2.2 Standard Precipitation INDEX (SPI) and Standard Precipitation Evapotranspiration Index (SPEI)

SPI index is calculated by taking the difference of the precipitation from the mean for a particular time scale, and then dividing it by the standard deviation (Table 9.1).

$$\text{SPI} = \frac{Xi - X}{\beta}$$

where Xi is mean precipitation for a particular time period, X is long-term mean precipitation, and β is standard deviation.

The Standard Precipitation Index (SPI), widely used in the region (Liu et al., 2016; Zhao et al., 2018; Sun et al., 2019; Wang et al., 2022; Dong et al., 2024), can be calculated for various time scales, providing early drought warnings and aiding

Table 9.1 Categories of SPI classification

SPI Values	Drought Category
2.00 +	Extremely wet
1.50 to 1.99	Very wet
1.00 to 1.49	Moderately wet
−0.99 to 0.99	Near normal
−1.00 to −1.49	Moderately dry
−1.50 to −1.99	Severely dry
−2.00 and less	Extremely dry

in the assessment of drought severity. However, SPI focuses solely on precipitation and does not account for soil moisture. In contrast, the Standardized Precipitation Evapotranspiration Index (SPEI) offers a more comprehensive assessment by incorporating evapotranspiration but requires more extensive data for calculation (Zhao et al., 2020; Han et al., 2021; Ge et al., 2022; Pei et al., 2022).

9.2.3 *The Palmer Drought Severity Index (PDSI)*

The Palmer Drought Severity Index (PDSI) was developed by W.C. Palmer as the first comprehensive drought index in the United States (Palmer, 1965). It is classified into different categories (Table 9.2).

The PDSI relies on a soil moisture algorithm calibrated for relatively homogeneous regions in the Midwest. Since the data used to develop the PDSI primarily originated from the Central Plains, this empirically derived index is less suitable for regions with diverse topographies, such as mountainous areas with significant elevation variations. Furthermore, the PDSI only accounts for rainfall as a form of precipitation, limiting its applicability in regions where snowfall serves as a major source of moisture. Despite these limitations, the PDSI has been extensively utilized by researchers studying the Yellow River basin (Zhang et al., 2013; Chen et al., 2014; Guo et al., 2021; Li et al., 2022).

9.2.4 *Crop Moisture Index (CMI)*

The Crop Moisture Index (CMI), a derivative of the Palmer Drought Severity Index, is designed to reflect short-term moisture supply across major crop-producing regions. Unlike the PDSI, it is not intended for assessing long-term drought conditions. Instead, it is primarily used to identify potential agricultural droughts, particularly in regions like the Loess Plateau (Zhang et al., 2017a, 2017b; 2020, Zhang et al., 2022a, 2022b); Xu et al., 2020.

Table 9.2 Categories of PDSI classification

PDSI Value	Category
≥ 4.0	Extremely wet
3.0 to 3.99	Very wet
2.0 to 2.99	Moderately wet
1.0 to 1.99	Slightly wet
0.5 to 0.99	Incipient wet spell
0.49 to −0.49	Near Normal
−0.5 to −0.99	Incipient dry spell
−1.0 to −1.99	Mild drought
−2.0 to −2.99	Moderate drought
−3.0 to −3.99	Severe drought
≤ −4.0	Extreme drought

9.2.5 *Surface Water Supply Index (SWSI)*

SWSI is based on snowpack, streamflow, precipitation, and reservoir storage. It represents water supply conditions unique to each basin, which limits interbasin comparisons. The index should be mainly used to assess the severity of hydrological drought, but it has been used in assessing the impact of drought on crop growth in the region as well (Li et al., 2020; He et al., 2020; Shi et al., 2022).

9.3 Temporal and Spatial Variation of Droughts

There have been numerous studies on the temporal and spatial variations of drought in the Yellow River basin, particularly in the middle reaches. Zhang et al. (2012) analyzed the droughts in the Loess Plateau (LP) over the period of 1970 to 2010 by combining the Variable Infiltration Capacity model with the Palmer Drought Index and found that the drought affected areas show an uptrend in the northwest and north of the LP with lower annual precipitation but a generally downtrend in the southeast and central area of the LP with a higher mean annual precipitation. Han et al. (2021) used standard soil moisture index to assess the drought frequency and coverage area in the Loess Plateau over the period of 1982–2015. Their results indicated that drought affected on average 10–25% of the LP annually and the drought frequency increases from southeast to northwest. Additionally, droughts occurring after 2000 have been less severe compared to those before 2000. Wu et al. (2018) studied 17 catchments in the LP and found that the region experienced a tendency towards both meteorological and hydrological droughts over the period of 1961 to 2013, with a hydrological drought more severe than meteorological drought. Liu et al. (2016) used the precipitation data collected from 54 meteorological stations across the LP and calculated the Standardized Precipitation Index (SPI) and the Standardized Precipitation Evapotranspiration Index (SPEI). The results showed that the drought frequency spatially decreased from southeast to northwest and was higher in the middle of the winter, late spring and early summer. Hou et al. (2021) based on the monthly precipitation and temperature to calculate the drought frequency from 1986 to 2019 and found that the trend of aridification became more significant in the Loess Plateau. The drought in different seasons shows difference in the Loess Plateau, with the highest frequency of drought in the spring and winter in the southeast and west of the Loess Plateau, whereas the drought in summer and autumn was more frequent in the northwest. The discrepancy in above research results result from the different study periods, and index and statistical methods that are employed in different studies.

9.4 Causes, Impacts and Mitigation of Drought

Drought in the region is directly influenced by temporal variations in meteorological factors. For example, in our long-term study site, the WJG watershed, the mean annual precipitation from 1955 to 2000 was 476 mm, ranging from 230 mm to 756 mm. Rainfall is highly seasonal, with 79% of the annual precipitation occurring between May and September. The coefficient of variation for summer precipitation is 0.365, significantly higher than the annual variation of 0.277. With most rainfall occurring during the summer months, soil moisture gradually decreases from autumn to spring, reaching its lowest levels in spring. This seasonal trend often results in frequent droughts from late spring to summer if expected rainfall does not occur. Rising temperatures and strong winds can further exacerbate drought severity.

Climate change can influence the severity and frequency of droughts in the region, a subject explored in numerous studies (Li et al., 2010; Wang et al., 2012a, 2012b; Sun et al., 2015). Most research indicates a clear warming trend in temperature over recent decades. However, findings on precipitation are mixed: some studies report a decrease in precipitation (Huang et al., 2009; Sun et al., 2015), while others observe no significant changes in the region (Wang et al., 2012a, 2012b).

The severity of drought is closely linked to soil moisture, which varies across different types of landuse. In the WJG watershed, we took soil moisture measurements over the summer season that revealed significant differences: check-dam land exhibited the highest soil moisture at 16.75%, compared to 8.52% on terrace land and 8.27% on slope cropland. Grassland and shrubland showed much lower average soil moisture levels, at 2.61% and 3.43%, respectively. The high soil moisture on check-dam land is attributed to its ability to retain runoff during storms, whereas the lower moisture levels on vegetated land are due to increased evapotranspiration and rainfall interception by plants (Wang et al., 2012a, 2012b). Declining soil moisture and reduced soil water storage have led to soil desiccation and the formation of deep dried soil layers in the reforestation land in the region (Jia & Shao, 2014; Yang et al., 2014; Jia et al., 2015). Interestingly, a study by Jia et al. (2017) found that after an initial significant decrease in soil moisture during the first 20–30 years after tree planting, soil moisture levels gradually begin to recover over time.

Several strategies can be implemented to combat drought and improve agricultural resilience in the region. First, check-dam land and terrace land should be expanded for food production. Check-dam land has proven to be significantly more productive, yielding 2–3 times more than terrace land and 6–10 times more than slope cropland, with even greater disparities observed during dry years. Despite accounting for only 9% of the total cropland area, check-dam land contributes 20.9% of the region's total food production (Wang et al., 2011). This highlights its importance in enhancing agricultural resilience and productivity in drought-prone areas. However, check-dam land is susceptible to flood damage, which can lead to reduced crop yields. Second, the cultivation of drought-resistant crops, such as millet and barley, should be prioritized in the drought-prone subregions of the northwest and north. Third, some agronomic measures such as returning mulch to crop fields can reduce evaporation, enhance soil moisture, and increase crop yields. For instance, combining straw mulch with irrigation has been shown to boost wheat yields by 52% during very dry years (Huang et al., 2005). Fourth, irrigation efficiency in irrigated flatlands, such as the Hetao Irrigation District and Yinchuan Irrigation District, should be further improved. In the Yinchuan Plain alone, approximately 4 billion tons of water are diverted from the Yellow River for irrigation annually (Li et al., 2018). The Hetao Irrigation District, the largest self-flowing irrigation district in Asia, covers an irrigation area of 1.162 million hectares (Zhang et al., 2017a, 2017b; Zhao et al., 2023). Currently, crop production in these districts relies on flood irrigation, or surface irrigation, which diverts water directly from the Yellow River. Transitioning from flood irrigation to drip irrigation could significantly improve water use efficiency and address soil salinization issues.

Drought also adversely affects revegetation efforts, particularly large-scale tree-planting. Artificial afforestation should be suitable for the local climate conditions. There is still ongoing debate about the suitability of afforestation in the Loess Plateau. Some argued that a large-scale afforestation is not suitable and sustainable in the Loess Plateau (Sun et al., 2022). Zhang et al. (2018) indicated that the current vegetation cover has exceeded the climate-defined equilibrium vegetation cover in many parts of the Loess Plateau. Others suggest that afforestation should be implemented in the more humid counties with low vegetation cover (Wu et al., 2021). The Loess Plateau has a great variation in mean precipitation, from less than 200 mm in the north and northwest to over 650 mm in the southeast, revegetation should be adapted to this spatial variability of rainfall conditions. Generally, afforestation is more appropriate in the more humid area while grass is more suitable to drier areas. However, even within a small watershed, the suitability of revegetation should also accommodate to the topographic and soil water conditions. With respect to tree planting, less water consumption tree species should be chosen. Drought can also impact pastures in the upper and middle reaches, potentially leading to grassland deterioration (Zhou et al., 2005). Therefore, implementing effective grassland management practices is crucial.

9.5 Salinization and Mitigation

According to the analysis of remote sensing data by Hong et al. (2013), salinization area accounts for 8.14% of the basin area, among which severe and extreme severe salinization occupy 14,548 km^2 and 4372 km^2, or 1.81% and 0.55% of the basin area, respectively. The most extensive and severe salinization areas are concentrated in the Yellow River delta and the irrigation districts. The salinization in the Yellow River delta is caused by both natural and anthropogenic factors. The unique geographic location of interzone between freshwater and saltwater is prone to seawater recharge. The severity of the salinization shows a decrease trend with the distance to the Bohai Sea. The decreasing flow discharge of Yellow River makes the seawater intrusion even more frequent. The low-lying ground, shallow ground water containing high content of salts, and semi-humid with strong seasonal variation of precipitation and temperature enhance evaporation and concentration of salts in the topsoil (Fan et al., 2012; Fu et al., 2020; Yang et al., 2017; Guo et al., 2019; Guo et al., 2023). The reclamation of coastal areas for agriculture and inappropriate land management led to secondary salinization. Oil exploitation and saline aquaculture can further deteriorate salinization (Zhang et al., 2011). Salinization in Hetao Plain and Yinchuan Plain is directly related to irrigation which led to the rise of ground water table. The area, located in the transition zone from semi-arid to arid climate, has strong evaporation that enhances the upward movement of ground water and increases the salinity of topsoils (Wang et al., 2016). Nevertheless, the spatial variation of salinization exists due to the difference in distance to the river, sedimentary environment, hydrogeological conditions and hydrology (Dong et al., 2022). Wang et al. (2023a, 2023b) found that the soil salinity in the Hetao Irrigation District increases with the distance along the river, which is ascribed to the progressively slower flow of the river system, the higher salt concentration of the irrigation water, and the shallower groundwater table. Salinization can also develop in the lands behind the check dams in the Loess Plateau region, which results from the interplay between a near-surface water table, a high evaporation rate, landform features, and the alkaline properties of the loessial soil (Jin et al., 2019). In the upper reaches of the Yellow River, Qinghai Plateau, salinization is largely limited to the area adjacent to the brackish lakes (Hong et al., 2023), though several studies indicated that the dynamic processes taking place in the permafrost layers can develop saline soils as well (Luo et al., 2020). The presence of permafrost prevents the downward soil drainage that results in the evaporative concentration of soluble salts in the thawing layers with the degradation of permafrost due to an increase in temperature.

Soil salinization has adverse impacts on crop yield and plant growth due to osmotic stress, salt stress and nutritional imbalance. Xie et al. (2021) indicated that the saline soils are not suitable for planting food crops if the salinity is above 3.5‰ whereas the net economic benefit of crops was negative under 3.5‰ salinity along the coast areas. Studies conducted elsewhere indicated that there is a linear decrease of crop yield with an increase in salinity (Hanks et al., 1978; Maas & Grattan, 1999). Soil salinization can cause significant damage to vegetation covers, soil erosion and land degradation (Chen et al., 2019).

Different mitigation strategies have been adopted in alleviating soil salinization aiming for the agricultural sustainability, such as drainage engineering measures, planting salt tolerating crops and plants, as well as chemical measures. Since 2013, the Ministry of Science and Technology of China launched a so-called "Bohai Granary" project in the Yellow River delta, aimed at solving saline soil agricultural production problem. The implementation of various strategies to manage saline soils includes selecting salt-tolerant crops such as rice and soybeans (Chen et al., 2013; Liu et al., 2022), adopting coastal grass-belt models to support livestock husbandry (e.g., forage rotation, rice-triticale/oat rotation, and forest-grass intercropping) (Wang et al., 2022), and utilizing agricultural engineering techniques such as field leveling and digging ditches to leach salts from the soil. Additionally, biological practices like planting salt-absorbing plants are also applied (Bian et al., 2021; Zhang et al., 2023). In areas affected by irrigation-induced salinization, primarily located in the Hetao and Yinchuan irrigation districts of the Yellow River basin, improving irrigation efficiency is crucial to addressing the salinization issue. Transitioning from traditional flood irrigation to drip irrigation can significantly reduce surface evaporation and subsurface water leakage. This, in turn, lowers the groundwater table and minimizes salt accumulation in the topsoil (Wang et al., 2023a, 2023b). Additionally, upgrading irrigation canal systems and adjusting cropping patterns are appropriate measures to alleviate salinization and promote sustainable agriculture in the region (Xu et al., 2010; Zhang et al., 2011). From a chemical perspective, applying gypsum—either to irrigation water or directly to the soil—can help reduce the sodium adsorption ratio (SAR) and exchangeable sodium percentage (ESP), lower soil pH, and release calcium into the soil, thereby improving soil structure and fertility (Li & Hui Qian, 2014; Song et al., 2023).

9.6 Summary and Conclusions

The Yellow River basin has semi-arid and arid climates, making it susceptible to drought. While natural variability of precipitation is the primary cause of drought, human activities can exacerbate its severity. Due to the lack of consensus among researchers despite numerous studies on the future impact of climate change on drought, long-term mitigation strategies for adapting to climate change are highly challenging. Drought significantly affects agriculture in the basin, particularly in the rain-fed agricultural areas of the Loess Plateau. To mitigate drought, emphasis should be placed on changing land use patterns by converting slope land into terraces and constructing check-dam lands. These changes can be combined with agronomic measures to enhance soil moisture retention and reduce evaporation. To address salinization issues in the Yellow River basin, it is essential to implement targeted measures that tackle the root causes. These measures should incorporate one or more approaches, including engineering, agronomic, biological, and chemical methods.

References

Bian, L., Wang, J., Liu, J., & Han, B. (2021). Spatiotemporal changes of soil salinization in the Yellow River Delta of China from 2015 to 2019. *Sustainability, 13*(2), 822.

Chen, P., Yan, K., Shao, H., & Zhao, S. (2013). Physiological mechanisms for high salt tolerance in wild soybean (Glycine soja) from Yellow River Delta, China: Photosynthesis, osmotic regulation, ion flux and antioxidant capacity. *PLoS One, 8*(12), e83227.

Chen, F., Yuan, Y., Zhang, R., & Qin, L. (2014). A tree-ring based drought reconstruction (AD 1760–2010) for the loess plateau and its possible driving mechanisms. *Global and Planetary Change, 122*, 82–88.

Chen, H., Zhao, G., Li, Y., Wang, D., & Ma, Y. (2019). Monitoring the seasonal dynamics of soil salinization in the Yellow River delta of China using Landsat data. *Natural Hazards and Earth System Sciences, 19*(7), 1499–1508.

Dong, Q., Li, J., Cheng, Y., Ren, Y., Zhang, D., Wang, D., Sun, X., & Cao, W. (2022). Distribution characteristics and formation mechanisms of highly mineralized groundwater in the Hetao plain, Inner Mongolia. *Water, 14*(20), 3247.

Dong, Q., Zhu, L., Duan, Z., Wang, L., Chen, C., Li, Y., Zhu, W., & Gurung, S. B. (2024). Interdecadal variation of drought patterns and associated influencing factors in the north-south transitional zone of China. *Journal of Geographical Sciences, 34*(7), 1253–1279.

Fan, X., Pedroli, B., Liu, G., Liu, Q., Liu, H., & Shu, L. (2012). Soil salinity development in the yellow river delta in relation to groundwater dynamics. *Land Degradation & Development, 23*(2), 175–189.

Fu, Z., Wang, P., Sun, J., Lu, Z., Yang, H., Liu, J., Xia, J., & Li, T. (2020). Composition, seasonal variation, and salinization characteristics of soil salinity in the Chenier Island of the Yellow River Delta. *Global Ecology and Conservation, 24*, e01318.

Ge, C., Sun, S., Yao, R., Sun, P., Li, M., & Bian, Y. (2022). Long-term vegetation phenology changes and response to multi-scale meteorological drought on the loess plateau, China. *Journal of Hydrology, 614*, 128605.

Guo, B., Yang, F., Han, B., Fan, Y., Chen, S., Yang, W., & Jiang, L. (2019). A model for the rapid monitoring of soil salinization in the Yellow River Delta using Landsat 8 OLI imagery based on VI-SI feature space. *Remote Sensing Letters, 10*(8), 796–805.

Guo, M., She, D., Zhang, L., Li, L., Yang, Z. L., & Hong, S. (2021). Attribution of trends in meteorological drought during 1960–2016 over the loess plateau, China. *Journal of Geographical Sciences, 31*(8), 1123–1139.

Guo, B., Lu, M., Fan, Y., Wu, H., Yang, Y., & Wang, C. (2023). A novel remote sensing monitoring index of salinization based on three-dimensional feature space model and its application in the Yellow River Delta of China. *Geomatics, Natural Hazards and Risk, 14*(1), 95–116.

Han, Z., Huang, Q., Huang, S., Leng, G., Bai, Q., Liang, H., Wang, L., Zhao, J., & Fang, W. (2021). Spatial-temporal dynamics of agricultural drought in the loess plateau under a changing environment: Characteristics and potential influencing factors. *Agricultural Water Management, 244*, 106540.

Hanks, R. J., Ashcroft, G. L., Rasmussen, V. P., & Wilson, G. D. (1978). Corn production as influenced by irrigation and salinity—Utah studies. *Irrigation Science, 1*, 47–59.

He, Z., Jia, G., Liu, Z., Zhang, Z., Yu, X., & Xiao, P. (2020). Field studies on the influence of rainfall intensity, vegetation cover and slope length on soil moisture infiltration on typical watersheds of the loess plateau, China. *Hydrological Processes, 34*(25), 4904–4919.

Hong, M. M., Wang, J. L., & Han, B. M. (2023). Spatial-temporal variation and attribution of salinization in the Yellow River Basin from 2015 to 2020. *International Journal of Digital Earth, 16*(1), 446–463.

Hou, Q. Q., Pei, T. T., Chen, Y., Ji, Z. X., & Xie, B. P. (2021). Variations of drought and its trend in the loess plateau from 1986 to 2019. *Ying Yong sheng tai xue bao= The Journal of Applied Ecology, 32*(2), 649–660.

Huang, Y., Chen, L., Fu, B., Huang, Z., & Gong, J. (2005). The wheat yields and water-use efficiency in the loess plateau: Straw mulch and irrigation effects. *Agricultural Water Management, 72*(3), 209–222.

Huang, Y., Cai, J., Yin, H., & Cai, M. (2009). Correlation of precipitation to temperature variation in the Huanghe River (Yellow River) basin during 1957–2006. *Journal of Hydrology, 372*(1–4), 1–8.

Jia, Y. H., & Shao, M. A. (2014). Dynamics of deep soil moisture in response to vegetational restoration on the loess plateau of China. *Journal of Hydrology, 519*, 523–531.

Jia, X., Shao, M. A., Zhang, C., & Zhao, C. (2015). Regional temporal persistence of dried soil layer along south–north transect of the loess plateau, China. *Journal of Hydrology, 528*, 152–160.

Jia, X., Zhu, Y., & Luo, Y. (2017). Soil moisture decline due to afforestation across the loess plateau, China. *Journal of Hydrology, 546*, 113–122.

Jin, Z., Guo, L., Wang, Y., Yu, Y., Lin, H., Chen, Y., Chu, G., Zhang, J., & Zhang, N. (2019). Valley reshaping and damming induce water table rise and soil salinization on the Chinese loess plateau. *Geoderma, 339*, 115–125.

Li, P., & Hui Qian, J. W. (2014). Hydrogeochemistry and quality assessment of shallow groundwater in the southern part of the Yellow River alluvial plain (Zhongwei section), Northwest China. *Earth Sciences Research Journal, 18*(1), 27–38.

Li, Z., Zheng, F. L., Liu, W. Z., & Flanagan, D. C. (2010). Spatial distribution and temporal trends of extreme temperature and precipitation events on the loess plateau of China during 1961–2007. *Quaternary International, 226*(1–2), 92–100.

Li, P., Qian, H., & Wu, J. (2018). Conjunctive use of groundwater and surface water to reduce soil salinization in the Yinchuan plain, North-West China. *International Journal of Water Resources Development, 34*(3), 337–353.

Li, Y., Luo, L., Chang, J., Wang, Y., Guo, A., Fan, J., & Liu, Q. (2020). Hydrological drought evolution with a nonlinear joint index in regions with significant changes in underlying surface. *Journal of Hydrology, 585*, 124794.

Li, Q., Deng, Y., Wang, S., Gao, L., & Gou, X. (2022). A half-millennium perspective on recent drying in the eastern Chinese loess plateau. *Catena, 212*, 106087.

Liu, Z., Wang, Y., Shao, M., Jia, X., & Li, X. (2016). Spatiotemporal analysis of multiscalar drought characteristics across the loess plateau of China. *Journal of Hydrology, 534*, 281–299.

Liu, C., Mao, B., Yuan, D., Chu, C., & Duan, M. (2022). Salt tolerance in rice: Physiological responses and molecular mechanisms. *The Crop Journal, 10*(1), 13–25.

Luo, D., Jin, H., Bense, V. F., Jin, X., & Li, X. (2020). Hydrothermal processes of near-surface warm permafrost in response to strong precipitation events in the headwater area of the Yellow River, Tibetan Plateau. *Geoderma, 376*, 114531.

Maas, E. V., & Grattan, S. R. (1999). Crop yields as affected by salinity. *Agricultural Drainage, 38*, 55–108.

Palmer, W. C. (1965). *Meteorological drought. Research paper no. 45, U.S* (p. 58). Weather Bureau.

Pei, T., Hou, Q., Chen, Y., Ji, Z., Wu, H., Xie, B., Qi, P., & Zhang, J. (2022). Vegetation in arid areas of the loess plateau showed more sensitivity of water-use efficiency to seasonal drought. *Forests, 13*(5), 634.

Shi, X., Ding, H., Wu, M., Shi, M., Chen, F., Li, Y., & Yang, Y. (2022). A comprehensive drought monitoring method integrating multi-source data. *PeerJ, 10*, e13560.

Song, C., Song, J., Wu, Q., Shen, X., Hu, Y., Hu, C., Li, W., & Wang, Z. (2023). Effects of applying river sediment with irrigation water on salinity leaching during wheat-maize rotation in the Yellow River Delta. *Agricultural Water Management, 276*, 108032.

Sun, Q., Miao, C., Duan, Q., & Wang, Y. (2015). Temperature and precipitation changes over the loess plateau between 1961 and 2011, based on high-density gauge observations. *Global and Planetary Change, 132*, 1–10.

Sun, C. X., Huang, G. H., Fan, Y., Zhou, X., Lu, C., & Wang, X. Q. (2019). Drought occurring with hot extremes: Changes under future climate change on loess plateau, China. *Earth's Future, 7*(6), 587–604.

Sun, J., Li, G., Zhang, Y., Qin, W., & Wang, M. (2022). Identification of priority areas for afforestation in the loess plateau region of China. *Ecological Indicators, 140*, 108998.

Tang, X., Miao, C., Xi, Y., Duan, Q., Lei, X., & Li, H. (2018). Analysis of precipitation characteristics on the loess plateau between 1965 and 2014, based on high-density gauge observations. *Atmospheric Research, 213*, 264–274.

Wang, Y., Fu, B., Chen, L., Lu, Y., & Gao, Y. (2011). Check dam in the loess plateau of China: Engineering for environmental services and food security. *Environmental Science & Technology, 45*, 1.

Wang, S., Fu, B. J., Gao, G. Y., Yao, X. L., & Zhou, J. (2012a). Soil moisture and evapotranspiration of different land cover types in the loess plateau, China. *Hydrology and Earth System Sciences, 16*(8), 2883–2892.

Wang, Q. X., Fan, X. H., Qin, Z. D., & Wang, M. B. (2012b). Change trends of temperature and precipitation in the loess plateau region of China, 1961–2010. *Global and Planetary Change, 92*, 138–147.

Wang, J., Wu, J., & Jia, H. (2016). Analysis of spatial variation of soil salinization using a hydrochemical and stable isotopic method in a semiarid irrigated basin, Hetao plain, Inner Mongolia, North China. *Environmental Processes, 3*, 723–733.

Wang, T. T., Cao, L. W., Liu, Z. Q., Yang, Q. S., Chen, L., Chen, M., & Jing, H. C. (2022). Basic biology of forage grass for constructing coastal Grass Belt in Yellow River Delta. *Chinese Bulletin of Botany, 57*(6), 837–847.

Wang, Y., Gao, M., Chen, H., Fu, X., Wang, L., & Wang, R. (2023a). Soil moisture and salinity dynamics of drip irrigation in saline-alkali soil of Yellow River basin. *Frontiers in Environmental Science, 11*, 345.

Wang, X., Zhang, H., Zhang, Z., Zhang, C., Zhang, K., Pang, H., Bell, S. M., Li, Y., & Chen, J. (2023b). Reinforced soil salinization with distance along the river: A case study of the Yellow River Basin. *Agricultural Water Management, 279*, 108184.

Wu, J., Miao, C., Zheng, H., Duan, Q., Lei, X., & Li, H. (2018). Meteorological and hydrological drought on the loess plateau, China: Evolutionary characteristics, impact, and propagation. *Journal of Geophysical Research: Atmospheres, 123*(20), 11–569.

Wu, X., Wang, S., Fu, B., & Liu, J. (2021). Spatial variation and influencing factors of the effectiveness of afforestation in China's loess plateau. *Science of the Total Environment, 771*, 144904.

Xie, H., Li, J., Zhang, Y., Xu, X., Wang, L., & Ouyang, Z. (2021). Evaluation of coastal farming under salinization and optimized fertilization strategies in China. *Science of the Total Environment, 797*, 149038.

Xu, X., Huang, G., Qu, Z., & Pereira, L. S. (2010). Assessing the groundwater dynamics and impacts of water saving in the Hetao Irrigation District, Yellow River basin. *Agricultural Water Management, 98*(2), 301–313.

Xu, H., Qu, Q., Wang, M., Li, P., Li, Y., Xue, S., & Liu, G. (2020). Soil organic carbon sequestration and its stability after vegetation restoration in the loess hilly region, China. *Land Degradation & Development, 31*(5), 568–580.

Yang, L., Wei, W., Chen, L., Chen, W., & Wang, J. (2014). Response of temporal variation of soil moisture to vegetation restoration in semi-arid loess plateau, China. *Catena, 115*, 123–133.

Yang, J., Liu, G., Li, J., Liu, H., & Li, B. (2017). Relationship between soil salinization and groundwater characteristics in the Yellow River Delta. *Acta Pedologica Sinica, 54*(6), 1377–1385.

Zhang, T. T., Zeng, S. L., Gao, Y., Ouyang, Z. T., Li, B., Fang, C. M., & Zhao, B. (2011). Assessing impact of land uses on land salinization in the Yellow River Delta, China using an integrated and spatial statistical model. *Land Use Policy, 28*(4), 857–866.

Zhang, B., Wu, P., Zhao, X., Wang, Y., Wang, J., & Shi, Y. (2012). Drought variation trends in different subregions of the Chinese loess plateau over the past four decades. *Agricultural Water Management, 115*, 167–177.

Zhang, B., Wu, P., Zhao, X., Wang, Y., Gao, X., & Cao, X. (2013). A drought hazard assessment index based on the VIC–PDSI model and its application on the loess plateau, China. *Theoretical and Applied Climatology, 114*, 125–138.

Zhang, M., Cheng, G., Feng, H., Sun, B., Zhao, Y., Chen, H., Chen, J., Dyck, M., Wang, X., Zhang, J., & Zhang, A. (2017a). Effects of straw and biochar amendments on aggregate stability, soil organic carbon, and enzyme activities in the loess plateau, China. *Environmental Science and Pollution Research, 24*, 10108–10120.

Zhang, H., Xiong, Y., Huang, G., Xu, X., & Huang, Q. (2017b). Effects of water stress on processing tomatoes yield, quality and water use efficiency with plastic mulched drip irrigation in sandy soil of the Hetao Irrigation District. *Agricultural Water Management, 179*, 205–214.

Zhang, S., Yang, D., Yang, Y., Piao, S., Yang, H., Lei, H., & Fu, B. (2018). Excessive afforestation and soil drying on China's loess plateau. *Journal of Geophysical Research: Biogeosciences, 123*(3), 923–935.

Zhang, Q., Jia, X., Wei, X., Shao, M., Li, T., & Yu, Q. (2020). Total soil organic carbon increases but becomes more labile after afforestation in China's loess plateau. *Forest Ecology and Management, 461*, 117911.

Zhang, F., Li, S., Yue, S., & Song, Q. (2022a). The effect of long-term soil surface mulching on SOC fractions and the carbon management index in a semiarid agroecosystem. *Soil and Tillage Research, 216*, 105233.

Zhang, X., Wang, M., Zhang, D., Zhang, Y., & Wang, X. (2022b). Increasing soil organic carbon pools and wheat yields by optimising tillage and fertilisation on the loess plateau in China. *European Journal of Soil Science, 73*(1), e13197.

Zhang, Z., Li, T., Shao, P., Sun, J., Xu, W., & Zhao, Y. (2023). Effects of short-term nitrogen addition on rhizosphere and bulk soil bacterial community structure of three halophytes in the Yellow River Delta. *Land Degradation & Development, 34*, 3281.

Zhao, W., Yu, X., Ma, H., Zhu, Q., Zhang, Y., Qin, W., Ai, N., & Wang, Y. (2015). Analysis of precipitation characteristics during 1957-2012 in the semi-arid loess plateau, China. *PloS One, 10*(11), e0141662.

Zhao, Q., Chen, Q., Jiao, M., Wu, P., Gao, X., Ma, M., & Hong, Y. (2018). The temporal-spatial characteristics of drought in the loess plateau using the remote-sensed TRMM precipitation data from 1998 to 2014. *Remote Sensing, 10*(6), 838.

Zhao, A., Yu, Q., Feng, L., Zhang, A., & Pei, T. (2020). Evaluating the cumulative and time-lag effects of drought on grassland vegetation: A case study in the Chinese loess plateau. *Journal of Environmental Management, 261*, 110214.

Zhao, Y., Shi, H., Miao, Q., Yang, S., Hu, Z., Hou, C., Yu, C., & Yan, Y. (2023). Analysis of spatial and temporal variability and coupling relationship of soil water and salt in cultivated and wasteland at Branch Canal scale in the Hetao Irrigation District. *Agronomy, 13*(9), 2367.

Zhou, H., Zhao, X., Tang, Y., Gu, S., & Zhou, L. (2005). Alpine grassland degradation and its control in the source region of the Yangtze and yellow Rivers, China. *Grassland Science, 51*(3), 191–203.

Chapter 10
Wetland Management

Contents

10.1 Introduction

Wetlands cover for about 6% of the world's land areas (Sahagian & Melack, 1998). They play important roles in preventing floods, improving water quality, and biochemical recycling (Mitsch et al., 2009). They also provide an indispensable habit for an abundant amount of biological species (Gibbs, 2000). Wetlands are an important carbon pools, and contain about 12% of global carbon pool, and thus play an important role in global carbon cycle and affect climate changes (IPCC, 1996; Chmura et al., 2003). According to the second national wetland inventory, there is a total of 3.91 million hectares of wetlands in the Yellow River basin, among which are 80.4%, 12.5% and 7.1% in the upper, middle and lower watershed, respectively (Sun et al., 2020). In this chapter, the ecological functions of those wetlands, the efforts and challenges of wetland management in the basin will be discussed.

T. Zhu, *Watershed Environmental Degradation and Sustainable Development*,
https://doi.org/10.1007/978-3-032-02448-0_10

10.2 Wetlands in the Source Region of Yellow River

The source region of the Yellow River contains 3.13 million hectares of wetlands, contributing to the Tibetan Plateau's extensive alpine wetland system—the largest of its kind globally—covering a total of 7.33 million hectares (Pan et al., 2011). They are extensively distributed across the region, situated at elevations ranging from 3000 to 5000 meters. According to their geomorphic locations, Gao (2016) classified those wetlands into alpine wetlands, piedmont wetlands, valley wetlands, riverine wetlands, floodplain wetlands, lacustrine wetlands, riverine wetlands and fluvial terrace wetlands.

The climate in the region is an alpine continental climate with an annual precipitation of 200–600 mm. The average annual temperature is below freezing point, with a short, wet summer and dry cold winter. The wetlands are primarily dominated by perennial grasses such as *Kobresia pygmaea* (C.B. Clarke), *Kobresia humilis* (C.A. Mey.) Serg., and *Kobresia tibetica* Maxim. These wetlands have relatively low productivity but support high biodiversity, with up to 30 species per square meter (Chen, 2007). They are also habitats for wild animals such as white-lipped deer (*Cervus albirostris*), Tibetan wild ass (*Equus kiang*), Tibetan gazelle (*Procapra picticaudata*) and blue sheep (*Pseudois nayaur*) (Li & Li, 2002).The alpine cold meadows under the permafrost show great variability due to the difference in elevation, with an active layer from less than 1 m to more than 4 m, and a thickness of permafrost from less than 30 m to more than 70 m (Wang et al., 2001a, 2001b). The low decomposition rate of plants also makes the alpine wetlands an important carbon pool (Babel et al., 2014; Brierley et al., 2022).

The wetland has undergone noticeable changes in recent times. Ma et al. (2022) found that the total wetland area in the region showed a decreasing trend of 7.08%, primarily due to a significant decline in marsh wetlands. However, river and lake wetlands saw substantial increases between 2000 and 2010, rising by 4.04% and 4.21%, respectively. Li et al. (2016) reported that Ruoergai Swamp (Zoige) at the eastern margin of the Qinghai in the Yellow River Source Zone, the world's largest plateau peat wetland, has experienced significant shrinkage and degradation since 1950s.

The wetland has been under threat from human activities. Overgrazing is the major anthropogenic disturbance to the alpine meadow wetlands (Song et al., 2009; Li et al., 2023). The alpine meadows support the livestock of yak and Tibet sheep. Overgrazing led to declines in soil fertility conditions and plant community biomass as well as the changes of the composition of species and the succession of biological community (Wang et al., 2001a, 2001b; Zhang et al., 2021). The meadow degradation was particularly severe in 1980s and 1990s, with the area of degraded vegetation increased from 24.5% in 1980s to 34.5% in 1990s (Wang et al., 2001a, 2001b). With the reduced vegetation cover, the degraded meadow is vulnerable to soil erosion and desertification. Once the bare ground in the grassland-meadow ecosystem is greater than 60%, extreme severe erosion occurs (Gao et al., 2020). However, one interesting finding is the composition of plant species tend to be more complex with the degradation of the meadow (Wang et al., 2009). In addition, the rodents such as rokor and pika in the area pose great threats to the alpine meadow, causing damages and disappearances of aboveground biomass and below ground roots, which can further change soil properties such as decreasing soil moisture content, increasing soil bulk density and permeability, and reducing soil nutrient, which in turn deter vegetation growth (Li et al., 2021; Shi et al., 2022).

The region is very sensitive to global warming (Tong et al., 2014; Meng et al., 2016). With a warming trend but not significant increase in precipitation as projected, the alpine meadow area may further shrink (Wang et al., 2008; Li et al., 2009; Duan et al., 2022). Climate change is likely to increase frequency and severity of droughts that will decrease soil water in the wet soils, which allows penetration of oxygen in the soil profile and increase decomposition of organic matter. The increasing temperature also leads to the thawing of permafrost, releasing carbon dioxide and methane to the atmosphere due to the breakdown of frozen plants and animals by bacteria. The storage of carbon would likely decrease in the future (Wang et al., 2022a, 2022b; Lin et al., 2022).

The restoration of the degraded alpine meadows is a challenging task. A combination of biological, engineering and ecological measures is required to restore the degraded wetlands. The major barriers to such efforts are funding, public environmental awareness and land ownership (Tian, 2014). The key to the restoration is to reduce the density of livestock, which is the main livelihood income source for local people. This requires a sustainable approach to environmental conservation and economic development.

10.3 Wetlands in the Middle Reaches of the Yellow River

The wetland area in the middle reaches of the Yellow River is limited today; however, studies indicate that around 4000 B.P., there was a significant extent of wetlands, particularly in the western Loess Plateau, due to wetter climatic conditions (An et al., 2005). With the drying trend of the climate, wetland has been shrunk significantly since then. Most of the current wetlands in the region belong to the riparian wetlands and floodplain wetlands. Overbank flooding and its temporal variability are the driving force for the dynamics of these wetlands. The reduction in river flow discharge, caused by decreased precipitation and increased human withdrawal, has led to a decline in the frequency and duration of overbank floods, posing significant threats to the health of riparian wetland ecosystems (Tian et al., 2010). In recent years, more artificial wetlands have been constructed on the Loess Plateau (Hao et al., 2019; Xiao, 2014). Reservoir construction also leads to the development of wetlands. However, due to hydrological changes, particularly a drop in water levels, reservoir wetlands have experienced a significant decline in recent years. For instance, the wetlands surrounding the Sanmenxia Reservoir decreased by 36.19% between 2000 and 2013 (Yang et al., 2020).

10.4 Wetlands in the Lower Reaches of the Yellow River

The wetlands in the lower reaches are mainly located in the river delta, with a total area of 4167 km^2 or 69.3% of the delta area (Wang et al., 2012). They can be classified into marsh wetland, meadow wetland, and tidal wetland (Li et al., 2009a).

The wetlands in the Yellow River delta support an abundant flora and fauna, with more than 200 plant species and more than 800 animal species (including 199 bird species) are identified in the wetlands (Cui et al., 2009). There are seven bird species listed as first priority, and 33 bird species listed as second priority for national conservation of China (Xu et al., 2004). The wetlands provide an exceptional habitat for breeding, migration and wintering. It is an important "transit station, overwintering place, and breeding place" for bird migration in the western Pacific Ocean of the inland river ring of Northeast Asia (Lu & Jiang, 2022).

Estuary wetlands, situated at the interface of land and sea, are heavily influenced by river flow and seawater dynamics. The flow discharge and sediment yield of the Yellow River play a crucial role in the formation and evolution of these wetlands. Since the 1970s, the Yellow River's flow discharge has steadily decreased due to a combination of natural and human factors. The construction of large reservoirs has altered the seasonal variation of flow discharges, while economic development in the basin has led to increased water consumption. The lower reaches of the Yellow River experienced annual dry-ups, with a peak of 226 dry days in 1997 (Xu et al., 2004). Although reservoir regulation has improved the situation since then, the annual water flow discharge into the sea continues to show a steady decline. The sediment discharge of the Yellow River also shows a drastic decrease largely due to the soil and conservation work implemented in the middle reaches of the Yellow River, with sediment load decreased from an average of 1.6 billion tons before 1950s to just a mere 0.4 billon tons in the 1990s. The decreased sediment deposition, together with the erosion caused by sea waves led to the changes of the delta development. The expansion of the delta decreased at the rate of 32.4 km^2 before 1979 to a 2.7 km^2 after then (Li et al., 2009a). Inappropriate human activities significantly contribute to wetland loss and degradation. Rapid population growth and urbanization in the delta region have accelerated the loss of wetlands. Dongying has grown into a major city with a population of 2.2 million, driven by the development of the petroleum industry. Over the period of 1976 to 2016, there was a net decrease of 103 km^2, including a loss of 954.03 km^2 of natural wetlands and an increase of 851.06 km^2 of artificial wetlands (Cong et al., 2019). The lost wetlands were converted to ponds and reservoirs, as well as farmlands. The wetlands are also seriously degraded due to pollutions, especially caused by oilwell blowouts and oil pipes leaking, as well as water extraction and soil salinization (Xu et al., 2004; Zhang & Sun, 2005; Huang et al., 2012). The delta region is the home to the Shengli Oil Field—the second largest in China. The expansion of oil infrastructure has resulted in wetland fragmentation and significant biodiversity loss.

To protect the wetlands, the Chinese government established a Yellow River Delta Nature Reserve in 1990. A range of conservation measures have been taken.

One of the key factors contributing to wetland loss and degradation in the Yellow River delta is the reduction in river water flow, which has led to increased seawater intrusion and coastal erosion by sea waves. To address this issue, since 1997, measures have been taken to increase water releases from reservoirs. These efforts aim to prevent the drying up of the lower river channel and maintain freshwater levels in the delta wetlands. Additionally, the construction of dikes along the coast has been implemented to prevent seawater erosion. To address soil salinity, salt-tolerant plant species have been introduced, and bioremediation techniques have been employed to remove pollutants from petroleum-contaminated soils (Yu et al., 2011; Xie et al., 2017; Wu et al., 2021; Wang et al., 2022a, 2022b). In addition, artificial wetlands have been developed to compensate for the loss of natural wetlands. Field monitoring following wetland restoration showed continuous improvements in water quality, biodiversity of flora and fauna, bird habitats, hydrological processes, and soil quality (Cui et al., 2009; Hua et al., 2012; Chen et al., 2016).

10.5 Conclusions

Wetlands in the Yellow River basin are primarily located in the source regions and the deltaic area. The source region wetlands, part of the world's largest alpine wetland, hold unique ecological value and play a significant role in biodiversity and carbon storage. However, they face serious threats from climate change and human activities, particularly overgrazing. Effective management of livestock grazing, alongside the establishment of wetland conservation measures, is essential for restoring these degraded wetlands.

At the mouth of the river, the deltaic wetlands—part of the world's youngest large river delta—are essential for biodiversity, providing crucial habitats for bird breeding and migration. However, these wetlands face significant threats from altered river flows, reduced sediment supply, land reclamation driven by rapid population growth and urbanization, and rising sea levels due to global warming. The establishment of a natural reserve in the region in recent years has been a vital step in protecting these wetlands. Moving forward, addressing these challenges will require regulating water discharge from reservoirs, along with a combination of biological and engineering solutions. Additionally, the creation of artificial wetlands can help offset the loss of natural wetlands and support ecological restoration.

References

An, C. B., Tang, L., Barton, L., & Chen, F. H. (2005). Climate change and cultural response around 4000 cal yr BP in the western part of Chinese loess plateau. *Quaternary Research, 63*(3), 347–352.

Babel, W., Biermann, T., Coners, H., et al. (2014). Pasture degradation modifies the water and carbon cycles of the Tibetan highlands. *Biogeosciences, 11*(23), 6633–6656.
Brierley, G., Li, X. L., Fryirs, K., Gao, J., Shi, Y., Perry, G. L., & Cullum, C. (2022). Degradation and recovery of alpine meadow catenas in the source zone of the Yellow River, Western China. *Journal of Mountain Science, 19*(9), 2487–2505.
Chen, G. C. (2007). In X. Q. Chen & X. J. Gou (Eds.), *Ecological environment in Sanjiangyuan natural reserve*. Qinghai People's Press. (In Chinese).
Chen, A., Sui, X., Wang, D., Liao, W., Ge, H., & Tao, J. (2016). Landscape and avifauna changes as an indicator of Yellow River Delta wetland restoration. *Ecological Engineering, 86*, 162–173.
Chmura, G. L., Anisfeld, S. C., Cahoon, D. R., & Lynch, J. C. (2003). Global carbon sequestration in tidal, saline wetland soils. *Global Biogeochemical Cycles, 17*(4), 1.
Cong, P., Chen, K., Qu, L., & Han, J. (2019). Dynamic changes in the wetland landscape pattern of the Yellow River Delta from 1976 to 2016 based on satellite data. *Chinese Geographical Science, 29*, 372–381.
Cui, B., Yang, Q., Yang, Z., & Zhang, K. (2009). Evaluating the ecological performance of wetland restoration in the Yellow River Delta, China. *Ecological Engineering, 35*(7), 1090–1103.
Duan, C., Li, X., Li, C., Yang, P., Shi, Y., Chai, Y., & Xu, W. (2022). Analysis on the soil physical, chemical, and microbial community properties of different alpine meadow patches in the source zone of the Yellow River, West China. *Ecological Indicators, 144*, 109531.
Gao, J. (2016). Wetland and its degradation in the Yellow River source zone. In *Landscape and ecosystem diversity, dynamics and management in the Yellow River Source Zone* (pp. 209–232). Springer.
Gao, X. X., Dong, S. K., Li, S., et al. (2020). Using the random forest model and validated MODIS with the field spectrometer measurement promote the accuracy of estimating aboveground biomass and coverage of alpine grasslands on the Qinghai-Tibetan plateau. *Ecological Indicators, 112*, 106114.
Gibbs, J. P. (2000). Wetland loss and biodiversity conservation. *Conservation Biology, 14*(1), 314–317.
Hao, H., Li, Y., Zhang, H., Zhai, R., & Liu, H. (2019). Spatiotemporal variations of vegetation and its determinants in the National key Ecological Function Area on loess plateau between 2000 and 2015. *Ecology and Evolution, 9*(10), 5810–5820.
Hua, Y., Cui, B., & He, W. (2012). Changes in water birds habitat suitability following wetland restoration in the Yellow River Delta, China. *Clean–Soil, Air Water, 40*(10), 1076–1084.
Huang, L., Bai, J., Chen, B., Zhang, K., Huang, C., & Liu, P. (2012). Two-decade wetland cultivation and its effects on soil properties in salt marshes in the Yellow River Delta, China. *Ecological Informatics, 10*, 49–55.
IPCC (International Panel on Climate Change). (1996). *Climate change 1996: Impacts, adaptations and mitigation of climate change: Scientific technical analysis. Contribution of working group II to the second assessment report of the IPCC*. Cambridge University Press.
Li, D., & Li, J. (2002). *Biodiversity of Sanjiangyuan—Scientific investigation report of Sanjiangyuan nature reserve, Beijing, China* (p. 4).
Li, F. X., Chang, G. G., Xiao, J. S., Zhou, B. R., & Fu, Y. (2009). Relationship between wetlands changes and climate change in the Yellow River source region. *Journal of Natural Resources, 24*(4), 683–690.
Li, S. N., Wang, G. X., Deng, W., Hu, Y. M., & Hu, W. W. (2009a). Influence of hydrology process on wetland landscape pattern: A case study in the Yellow River Delta. *Ecological Engineering, 35*(12), 1719–1726.
Li, Z., Wang, Z., & Pan, B. (2016). Wetland ecosystems of the Yellow River source zone. In *Landscape and ecosystem diversity, dynamics and management in the Yellow River Source Zone* (pp. 183–207). Springer.
Li, J., Li, X., Gao, J., Ma, G., & Qi, X. (2021). Micro-scale fragmentation of the alpine meadow landscape on the Qinghai-Tibet plateau under external disturbances. *Catena, 201*, 105220.

Li, X., Zhang, Y., Yan, H., & Salahou, M. K. (2023). Watershed-level spatial pattern of degraded alpine meadow and its key influencing factors in the Yellow River source zone of West China. *Ecological Indicators, 146*, 109865.

Lin, Q., Wang, S., Li, Y., Riaz, L., Yu, F., Yang, Q., Han, S., & Ma, J. (2022). Effects and mechanisms of land-types conversion on greenhouse gas emissions in the Yellow River floodplain wetland. *Science of the Total Environment, 813*, 152406.

Lu, M., & Jiang, Y. (2022). Research progress on the evolution characteristics of wetlands in the Yellow River Delta. *Advances in Resources Research, 2*(3), 87–98.

Ma, T., She, Y., Zhao, L., Hu, B., Feng, X., Zhao, J., & Zhao, Z. (2022). Alpine wetland evolution and their response to climate change in the yellow-river-source national park from 2000 to 2020. *Water, 14*(15), 2351.

Meng, F., Su, F., Yang, D., Tong, K., & Hao, Z. (2016). Impacts of recent climate change on the hydrology in the source region of the Yellow River basin. *Journal of Hydrology: Regional Studies, 6*, 66–81.

Mitsch, W. J., Gosselink, J. G., Zhang, L., & Anderson, C. J. (2009). *Wetland ecosystems*. John Wiley & Sons.

Pan, X., Zhang, J., Luo, Z., Mao, L., Jin, H., Guo, L., Bi, X., & Zhao, Y. (2011). Natural wetland in China. *African Journal of Environmental Science and Technology, 5*(1), 45–55.

Sahagian, D., & Melack, J. (1998). Global wetland distribution and functional characterization: Trace gases and the hydrologic cycle. *IGBP Report, 46*, 92.

Shi, Y., Gao, J., Li, X., Li, J., & Brierley, G. (2022). Effects of disturbances on aboveground biomass of alpine meadow in the Yellow River source zone, Western China. *Ecology and Evolution, 12*(3), e8640.

Song, X., Yang, G., Yan, C., Duan, H., Liu, G., & Zhu, Y. (2009). Driving forces behind land use and cover change in the Qinghai-Tibetan plateau: A case study of the source region of the Yellow River, Qinghai Province, China. *Environmental Earth Sciences, 59*, 793–801.

Sun, G., Zhang, M., & Lei, G. (2020). Wetland water bird biodiversity conservation strategies in the Yellow River basin. *Biodiversity Science, 28*(12), 1469–1482.

Tian, K. (2014). In International advisory board Chinese research perspectives online (Ed.), *Plateau wetlands: Functions, values, and environmental deterioration. Chinese research perspectives online*. Brill online.

Tian, S., Luo, W., Jia, Z., & Butalia, R. S. (2010). Temporal variation of over-bank flooding of Wei River and its impact on a riparian wetland in Xi'an, China. *Hydrological Processes: An International Journal, 24*(10), 1296–1307.

Tong, L., Xu, X., Fu, Y., & Li, S. (2014). Wetland changes and their responses to climate change in the "three-river headwaters" region of China since the 1990s. *Energies, 7*(4), 2515–2534.

Wang, G., Cheng, G. and Shen, Y., (2001a). Research on ecological environmental changes in the source regions of Yangtze–yellow Rivers and their integrated protection.

Wang, G., Qian, J., Cheng, G., & Lai, Y. (2001b). Eco-environmental degradation and causal analysis in the source region of the Yellow River. *Environmental Geology, 40*, 884–890.

Wang, G. X., Li, Y. S., Wang, Y. B., & Wu, Q. B. (2008). Effects of permafrost thawing on vegetation and soil carbon pool losses on the Qinghai–Tibet Plateau, China. *Geoderma, 143*, 143–152.

Wang, C. T., Long, R. J., Wang, Q. L., Jing, Z. C., & Shi, J. J. (2009). Changes in plant diversity, biomass and soil C, in alpine meadows at different degradation stages in the headwater region of three rivers, China. *Land Degradation & Development, 20*(2), 187–198.

Wang, M., Qi, S., & Zhang, X. (2012). Wetland loss and degradation in the Yellow River delta, Shandong Province of China. *Environmental Earth Sciences, 67*, 185–188.

Wang, Q., Jin, H., Wu, Q., Zhang, T., Yuan, Z., Li, X., Ming, J., Yang, C., Serban, R. D., & Huang, Y. (2022a). The vertical distribution of soil organic carbon and nitrogen in a permafrost-affected wetland on the Qinghai–Tibet Plateau: Implications for Holocene development and environmental change. *Permafrost and Periglacial Processes, 33*(3), 286–297.

Wang, B., Kuang, S., Shao, H., Cheng, F., & Wang, H. (2022b). Improving soil fertility by driving microbial community changes in saline soils of Yellow River Delta under petroleum pollution. *Journal of Environmental Management, 304*, 114265.

Wu, T., Li, X. B., Xu, J., Liu, L. X., Ren, L. L., Dong, B., Li, W., Xie, W. J., Yao, Z. G., Chen, Q. F., & Xia, J. B. (2021). Diversity and functional characteristics of endophytic bacteria from two grass species growing on an oil-contaminated site in the Yellow River Delta, China. *Science of The Total Environment, 767*, 144340.

Xiao, J. (2014). Satellite evidence for significant biophysical consequences of the "grain for green" program on the loess plateau in China. *Journal of Geophysical Research: Biogeosciences, 119*(12), 2261–2275.

Xie, W., Zhang, Y., Li, R., Yang, H., Wu, T., Zhao, L., & Lu, Z. (2017). The responses of two native plant species to soil petroleum contamination in the Yellow River Delta, China. *Environmental Science and Pollution Research, 24*, 24438–24446.

Xu, X., Lin, H., & Fu, Z. (2004). Probe into the method of regional ecological risk assessment—A case study of wetland in the Yellow River Delta in China. *Journal of Environmental Management, 70*(3), 253–262.

Yang, H., Wang, J., Xiao, W., Lu, F., Wang, Y., & JarsjÖ, J. (2020). Relationship between hydroclimatic variables and reservoir wetland landscape pattern indices: A case study of the Sanmenxia reservoir wetland on the Yellow River, China. *Journal of Earth System Science, 129*, 1–11.

Yu, S., Li, S., Tang, Y., & Wu, X. (2011). Succession of bacterial community along with the removal of heavy crude oil pollutants by multiple biostimulation treatments in the Yellow River Delta. *China. Journal of Environmental Sciences, 23*(9), 1533–1543.

Zhang, J. F., & Sun, Q. X. (2005). Causes of wetland degradation and ecological restoration in the Yellow River Delta region. *Forestry Studies in China, 7*, 15–18.

Zhang, X., Wang, G., Xue, B., Zhang, M., & Tan, Z. (2021). Dynamic landscapes and the driving forces in the Yellow River Delta wetland region in the past four decades. *Science of the Total Environment, 787*, 147644.

Chapter 11
Environmental Impacts of Mining and Oil Production

Contents

11.1 Introduction

The Yellow River basin is an important energy base and mineral reserve area in China. The upper and middle reaches are the key areas for fossil fuel production, primarily located in Inner Mongolia, Shanxi, Shaanxi and Qinghai province. This region contains a total of 150.5 billion tons of coal, accounting for 20% of China's total reserves, along with 1.98 billion tons of oil and 1180 billion cubic meters of natural gas. The Yellow River basin is rich in metal resources, including copper, gold, aluminum, and lead. Shanxi Province holds an aluminum reserve of 1.641 billion tons, accounting for 41.57% of China's total (Tan et al., 2023). Gansu Province is renowned for its production of copper, lead, and zinc, while the Yellow River delta serves as the base for Shengli Oilfield Inc., the second-largest oilfield in China. However, with economic growth, mining activities have emerged as a major environmental concern in the region. This chapter explores the key environmental impacts of mining and oil production, along with strategies for restoration.

T. Zhu, *Watershed Environmental Degradation and Sustainable Development*,
https://doi.org/10.1007/978-3-032-02448-0_11

11.2 Coal Mining

Coal mining in the region primarily involves opencast and underground methods. Opencast coal mining is predominantly concentrated in Shanxi, Inner Mongolia, and Shaanxi, with some activities in the source region as well (Wohlfart et al., 2017). This process involves extensive excavation and displacement of surface materials. The excavated soil is typically dumped nearby, where it becomes highly compacted due to the use of large-scale machinery. As a result, the soil develops poor structure, low nutrient content, and limited water-holding capacity (Zhen et al., 2019; Zhang & Ling, 2020). Mining activities permanently alter the surface landscape, removing or damaging vegetation. The exposed soil, lacking vegetation cover, becomes highly susceptible to both water and wind erosion.

Underground coal mining constitutes the majority of coal production in the region. On average, waste generated during the process—such as coal gangue, coal sludge, and fly ash—accounts for approximately 15% of the total coal output (Liu & Liu, 2010). Coal gangue, primarily composed of rocks and non-coal materials, is a waste byproduct generated during the extraction and processing of coal. In coal mining areas, topsoils often contain varying amounts of coal gangue, which can significantly influence soil properties such as bulk density, structure, water retention, water movement, and solute transport rates (Zhou et al., 2010). Coal sludge, a byproduct of the coal washing process, contains carcinogenic chemicals and toxic heavy metals such as arsenic, mercury, chromium, cadmium, boron, selenium, and nickel. Fly ash is predominantly generated during coal combustion. The coal washing process consumes vast amounts of water, a critical issue in regions with limited water resources due to their semi-arid and arid climates. The wastewater produced is laden with suspended sediments, acids, and heavy metals, posing a significant risk of contaminating both surface and groundwater. Analyses shows that groundwater in the Shenfu coal mining area is contaminated by F^-, NO_3^-, NO_2^-, NH_4^+, Fe and Mn in some locations, which may pose risks on human health. Li and Qian (2018) assessed groundwater quality in the Hongdunzi coal mine and concluded that the groundwater is not suitable for direct human consumption and irrigation due to the high concentrations of F^- and NO_3^-. Domestic-consumption water quality improvement devices and rainwater harvesting are suggested as the measures for human consumption and groundwater quality protection (Su et al., 2018; Wu et al., 2019).

Underground mining causes surface subsidence which is the dominant cause of surface structure damages in the mining area (Kang, 2008; Yang et al., 2022). Coal mining activities can also trigger geological hazards such as landslides. Ground fissures caused by subsidence are widespread in underground coal mining areas (Mi et al., 2022) and may serve as precursors to landslides. For instance, a deep-seated landslide was observed in the Wangjiagou watershed, Shanxi Province, within a coal mining region (Fig. 11.1). Unlike most landslides in the Loess Plateau, which are typically triggered by heavy rainstorms during the rainy season, this landslide occurred in the spring without any rainfall trigger. Before the landslide, parallel

Fig. 11.1 Landslide triggered by underground coal mining in the WJG watershed on the Loess Plateau

ground fissures were noted developing on the terraces. Research indicates that slow subsidence can continue even after underground mining operations cease, potentially triggering shallow landslides in coal mining areas (Yang et al., 2022). Loose loess materials from such slides may transform into mudflows during subsequent rainstorms. Surface subsidence alters soil structure, disrupts biological communities, and affects surface hydrology and groundwater levels. Soils near fissures often exhibit high porosity and low bulk density, making them susceptible to tunnel and gully formation. Additionally, ground fissures facilitate rainwater infiltration, increasing soil weight and exacerbating slope instability. These fissures also contribute to the loss of surface soil nutrients, further degrading the land productivity.

Coal mining areas often experience severe soil erosion due to vegetation removal, steepened slopes, and loose soil materials. Changes in land use within these areas further contribute to soil fertility decline and the abandonment of croplands. Additionally, large amounts of coal gangue are generated during mining and processing. The accumulation of coal gangue poses fire hazards due to spontaneous combustion triggered by the presence of unburned coal, pyrite (FeS_2), and oxygen, releasing toxic emissions such as SO_2, CO, H_2S, Hg, Pb, and benzopyrene into the atmosphere (Zhai et al., 2017). Even after mining activities cease, the degraded ecosystems in these areas may remain impaired for extended periods. Furthermore, coal combustion in the power plants and steel factories in the region contributes significantly to air pollution, releasing carbon dioxide (CO_2), sulfur dioxide (SO_2), nitrogen oxides (NOx), volatile organic compounds (VOCs), and various heavy metals, including mercury (Hg).

Coal mining activities also contribute to soil contamination. For instance, a study of the Pingshuo coal mine, the largest opencast mine in China located in the Loess Plateau, revealed that heavy metals such as As, Cd, Cr, Cu, Pb, and Hg in reclaimed cultivated land (ranging from 2 to 22 years post-reclamation) exceeded the background levels for Shanxi Province. However, all concentrations remained below the threshold values specified by Chinese soil guidelines (Grade II standards) (Liu et al., 2017). Similarly, research in the coal mining region of northern Shaanxi Province found that average concentrations of heavy metals such as Hg, Cd, Pb, and Cr were 2.03, 1.36, 1.11, and 1.23 times higher, respectively, than the provincial soil background values, though other heavy metal concentrations were lower than the background levels (Wang et al., 2022). The distribution of heavy metal contamination in soils exhibited moderate spatial variation depending on land use types (Zhang et al., 2021).

Various remediation measures have been taken in post coal mining areas (Peng & Bi, 2020). The reclaimed lands include cultivation and revegetation of the mined lands. Studies showed that field capacity and bulk density of 15- and 18-year reclaimed mine soils after planting crops were close to the levels of the natural soils in the 0–10 cm and 50–80 cm soil layers. However, soil organic matter and total nitrogen are difficult to recover even after years' cultivation (Liu et al., 2017). Mined areas can be restored through revegetation using trees, shrubs, and grasses. In the Loess Plateau of China, *Robinia pseudoacacia* is the predominant tree species used for reclamation in opencast coal mines. In monoculture forests of *Robinia pseudoacacia*, soil bulk density and pH decrease over the reclamation chronosequence, while soil organic matter, phosphorus (P), and nitrogen (N) increase significantly during the initial stages before stabilizing over a 30-year period. It takes approximately 10 years for disturbed soils to recover to levels comparable to undisturbed soils (Yuan et al., 2018). Soil microorganisms, which are essential for organic matter decomposition and nutrient cycling, play a critical role in this recovery process. Vegetation rehabilitation enhances bacterial richness and metabolism in the soil, leading to improved soil quality in mining subsidence areas (Luo et al., 2020).

In coal-mining subsidence areas, ground fissures can be filled, and subsided land can be repurposed for cropland or vegetation. Incorporating crop and plant residues into the soil can promote aggregate formation, aiding in the gradual recovery of soil physical and chemical properties following land reclamation. Experimental efforts have also explored using sediments from the Yellow River to fill subsidence areas (Hu et al., 2015).

Large amount of waste coal gangue produced by coal mining can be reused as cement-based construction materials, as opposed to the common practices of the coal gangue by spreading and land filling which occupy land and leach heavy metals to soils and ground water (Wu et al., 2017; Zhang & Ling, 2020; Qiu et al., 2021). Experimental studies were also conducted by mixing waste coal gangue with dispersive soils that can limit the dispersity of soil and changing dispersive soil into nondispersive soil, subsequently enhancing the water stability of dispersive soil (Zhao et al., 2023). Others suggest reutilization of waste coal gangue in underground backfilling mining (Ma et al., 2021). This can protect the overburden aquifer though may also cause the groundwater contamination concerns.

Geographic Information Systems (GIS) and remote sensing technologies have been used in assessment of geohazards and eco-environment quality, prediction of floor water inrush and as well as restoration planning in the coal mining areas (Wu et al., 2009; Wu et al., 2011; Yang et al., 2019, 2022; Hou et al., 2021).

11.3 Non-ferrous Metal Mining

The non-ferrous metal mining and smelting in the Yellow River basin is also important, such as gold, copper, aluminium and zinc in Gansu, Inner Mongolia, Shanxi, and Qinghai. Many of those processing plants are located along the rivers. The treated and untreated industrial wastewater discharge into the rivers, and the farmers use the water in those streams for irrigation of farmlands (Li et al., 2014; Zang et al., 2017; Nan & Zhao, 2000). This poses potential risk of soil contamination, and a concern of crops produced. Soil samples in the agricultural land in Dongdagou and Xidagou stream valleys of the Baiyin city of Gansu province where copper mining activities occurred for decades show that the mean concentrations of Cu, Zn, Cd and Pb exceed the soil background concentrations of the region, and a fraction of Cd and Pb in the corn grain exceeds the maximum levels of contaminants in foods in China. The spatial distribution pattern of Cd in corn grains was highly similar to that in sampled soils, which suggests that the sources of heavy metals in corn are from the soils (Zang et al., 2017). Wang et al. (2016) conducted the heavy metal contaminations in soil and crops in the same area, showing that Hg levels in 11 out of 26 sampled soils (42.3% of soil samples) exceeded Hg limit of grade II soil environmental quality standards in China (1.0 $mg \cdot kg^{-1}$). Hg in over 50% of wheat grain samples reached or exceeded the maximum permissible food safety levels (0.02 $mg \cdot kg^{-1}$) according to the General Standard of Contaminants in Food in China (GB 2762–2012). In the Xiaoqinlin Gold Belt, the second largest gold mining area of China, soils are most highly enriched in Pb, followed by Cr, Cu, and Zn. As for the crops (corn and wheat), Zn and Cu are the most abundant elements, followed by Pb and Cr. Wheat is more liable to be enriched in heavy metals than corn is grown in all types of soils (Liu et al., 2020). There are various techniques available for restoring soils contaminated with heavy metals. These include phytoremediation, which utilizes plants to absorb and transform heavy metals in the soil (Salt et al., 1998; Pilon-Smits, 2005), and microbial remediation, which employs microbes to immobilize or alter heavy metals (Jin et al., 2018; Wang et al., 2021). Other methods include soil capping, which involves covering contaminated soil with a layer of clean soil, and soil excavation, which entails the physical removal of contaminated soil, among other approaches (Liu et al., 2018; Nguyen et al., 2022).

11.4 Oil Production

The Yellow River basin is important in petroleum production in China, including three of the top ten oil fields in China-Yanchang and Changqing in the middle reaches of Yellow River and Shenli in the Yellow River delta. The extraction, refinery, storage, and transport of petroleum may pose serious impacts on the environment. Petroleum contaminants include petroleum hydrocarbon, n-alkanes, polycyclic aromatic hydrocarbons (PAHs), resins, asphaltenes, among others. Petroleum can contaminate soils and water, threatening ecosystems. Crude oil, once leaking into soils, can change soil physical and chemical properties. It can cover soil aggregates and block soil structural inter-aggregates pores, leading to an anaerobic environment and further affecting the soil microbial communities (Gao et al., 2022). It can further enter food chain, posing great risk to human health. Surface soil and river sediment samples collected from the waste land, farmland, and riverbed in the Yanchang oilfield area show that the total petroleum hydrocarbon (TPH) and polycyclic aromatic hydrocarbons (PAHs) were 907–3447 mg/kg and 103.59–563.50 μg/kg, respectively, significantly higher than the control samples (TPH 224 mg/kg, PAHs below method quantification limit) (Wang et al., 2018). TPH values in another oil field area, Changqin, range 1678–6748, 1189–2237, and 1089–1728 mg/kg in the wastelands, sediments, and farmlands, respectively (Gao et al., 2019). The petroleum contaminants enter surface water and groundwater, leading to deterioration of water quality. A study conducted in the Longdong oilfield of the Loess Plateau shows that petroleum contamination caused by the oil industry is the largest source of pollution in the Malian river, leading to much of the water unsafe to drink and unsuitable for use in irrigation (Ma et al., 2012).

In the Yellow River delta where Shenli Oilfield is located, oil exploration has been carried out since 1960s, which could affect soil, vegetation and wetland ecosystem (Wang et al., 2012; Miao et al., 2020; Fu et al., 2022; Yuan et al., 2022). Field sediment samples collected from the sites of spill petroleum-hydrocarbon spillage showed that the concentrations of heavy metals, especially Cr, Cu, Mn, Pb, and Zn, are significantly higher than in unpolluted sediments (Nie et al., 2010). However, Miao et al. (2020) indicated that only Cd, As and Pb instead of all heavy metals in the soils are significantly elevated and affected by oil production and the concentration of other heavy metals affected by oil production in the Yellow River delta is relatively small. As heavy metals are highly resistant to degradation in the natural ecosystems, and thus they will be subject to bioaccumulation, which will pose a great threat to microbiota, flora and fauna. Soil contamination by petroleum hydrocarbons significantly impacts bacterial biodiversity (Gao et al., 2022). However, Wang et al. (2011) reported that contamination levels in the delta region are relatively low to moderate compared to other locations globally that are chronically affected by oil pollution. In the oil-contaminated soils, the medium (C15–C28) and heavy (C29–C38) fractions were identified as the dominant total petroleum hydrocarbon (TPH) components, while the light fraction (C10–C14) was negligible.

Additionally, the study suggested that polycyclic aromatic hydrocarbons (PAHs) currently do not pose adverse biological effects in the delta area.

The remediation measures of oil-contaminated soils include biological, physical and chemical, and integrated approaches (Lim et al., 2016). For example, biostimulation treatments were experimented to enhance the removal of heavy crude oil pollutants in the saline soil of Yellow River Delta (Yu et al., 2011); Local wild plant species were tested in the oil contaminated soils to assess the changes of soil bacteria and the effectiveness of total petroleum hydrocarbons removal (Xu et al., 2013). Nevertheless, most of those biological measures are in the experimental stage, and not applied to the field in a large scale.

11.5 Summary and Conclusions

The Yellow River basin is a major source of fossil fuels and mineral reserves, with mining being one of the region's primary economic activities. However, these activities have caused significant ecological and environmental degradation. In the past, numerous small coal mines operated in the Loess Plateau region, prioritizing short-term economic gains. Traditional mining techniques were employed, with minimal or no regard for environmental protection. As a result, the closure of these mines left severe damage to the surface and groundwater, soils, and vegetation. Restoring the degraded ecosystems presents challenges such as limited funding, technological constraints, and policy gaps. Moving forward, comprehensive environmental protection should be more strictly implemented. Mining companies should also be required to establish funding reserves beforehand for ecosystem restoration, as remediation costs are directly linked to the extent of environmental damage caused by mining. This approach would not only address the financial burden of restoration but also encourage environmental responsibility among mining companies. Remediating contaminated sites is a complex process that requires long-term efforts to restore soil, water, and biological communities. The process should start with an assessment to determine the severity and extent of the damage caused by mining or oil extraction. Next, a comprehensive restoration plan should be developed, detailing specific objectives, methods, and timelines. An integrated approach, utilizing various contamination cleanup techniques, may be essential to fully restore the integrity of the soil, water, and ecosystems in the affected mining areas.

References

Fu, Z., Ge, X., Gao, Y., Liu, J., Ma, Y., Yang, X., & Meng, F. (2022). Effects of salinity and oil contamination on the soil seed banks of three dominant vegetation communities in the coastal wetland of the Yellow River Delta. *Forests, 13*(4), 615.

Gao, S., Liang, J., Teng, T., & Zhang, M. (2019). Petroleum contamination evaluation and bacterial community distribution in a historic oilfield located in loess plateau in China. *Applied Soil Ecology, 136*, 30–42.

Gao, Y., Yuan, L., Du, J., Wang, H., Yang, X., Duan, L., Zheng, L., Bahar, M. M., Zhao, Q., Zhang, W., & Liu, Y. (2022). Bacterial community profile of the crude oil-contaminated saline soil in the Yellow River Delta natural reserve, China. *Chemosphere, 289*, 133207.

Hou, H., Ding, Z., Zhang, S., Guo, S., Yang, Y., Chen, Z., Mi, J., & Wang, X. (2021). Spatial estimate of ecological and environmental damage in an underground coal mining area on the loess plateau: Implications for planning restoration interventions. *Journal of Cleaner Production, 287*, 125061.

Hu, Z., Wang, P., & Shao, F. (2015). Technique for filling reclamation of mining subsidence land with Yellow River sediment. *Transactions of the Chinese Society of Agricultural Engineering, 31*(3), 288–295.

Jin, Y., Luan, Y., Ning, Y., & Wang, L. (2018). Effects and mechanisms of microbial remediation of heavy metals in soil: A critical review. *Applied Sciences, 8*(8), 1336.

Kang, J. R. (2008). Analysis of effect of fissures caused by underground mining on ground movement and deformation. *Chinese Journal of Rock Mechanics and Engineering, 27*(1), 59–64.

Li, P., & Qian, H. (2018). Water in loess. In R. A. Meyers (Ed.), *Encyclopedia of sustainability science and technology* (pp. 1–17). Springer.

Li, Z., Ma, Z., van der Kuijp, T. J., Yuan, Z., & Huang, L. (2014). A review of soil heavy metal pollution from mines in China: Pollution and health risk assessment. *Science of the Total Environment, 468*, 843–853.

Lim, M. W., Von Lau, E., & Poh, P. E. (2016). A comprehensive guide of remediation technologies for oil contaminated soil—Present works and future directions. *Marine Pollution Bulletin, 109*(1), 14–45.

Liu, H. B., & Liu, Z. L. (2010). Recycling utilization patterns of coal mining waste in China. *Resources, Conservation and Recycling, 54*(12), 1331–1340.

Liu, X., Bai, Z., Zhou, W., Cao, Y., & Zhang, G. (2017). Changes in soil properties in the soil profile after mining and reclamation in an opencast coal mine on the loess plateau, China. *Ecological Engineering, 98*, 228–239.

Liu, L., Li, W., Song, W., & Guo, M. (2018). Remediation techniques for heavy metal-contaminated soils: Principles and applicability. *Science of the Total Environment, 633*, 206–219.

Liu, R. P., Xu, Y. N., Zhang, J. H., Wang, W. K., & Elwardany, R. M. (2020). Effects of heavy metal pollution on farmland soils and crops: A case study of the Xiaoqinling Gold Belt, China. *China Geology, 3*(3), 402–410.

Luo, Z., Ma, J., Chen, F., Li, X., Zhang, Q., & Yang, Y. (2020). Adaptive development of soil bacterial communities to ecological processes caused by mining activities in the loess plateau, China. *Microorganisms, 8*(4), 477.

Ma, J., Pan, F., He, J., Chen, L., Fu, S., & Jia, B. (2012). Petroleum pollution and evolution of water quality in the Malian River basin of the Longdong loess plateau, northwestern China. *Environmental Earth Sciences, 66*, 1769–1782.

Ma, D., Zhang, J., Duan, H., Huang, Y., Li, M., Sun, Q., & Zhou, N. (2021). Reutilization of gangue wastes in underground backfilling mining: Overburden aquifer protection. *Chemosphere, 264*, 128400.

Mi, J., Yang, Y., Hou, H., Zhang, S., Ding, Z., & Hua, Y. (2022). Impacts of ground fissures on soil properties in an underground mining area on the loess plateau, China. *Land, 11*(2), 162.

Miao, X., Hao, Y., Zhang, F., Zou, S., Ye, S., & Xie, Z. (2020). Spatial distribution of heavy metals and their potential sources in the soil of Yellow River Delta: A traditional oil field in China. *Environmental Geochemistry and Health, 42*, 7–26.

Nan, Z., & Zhao, C. (2000). Heavy metal concentrations in gray calcareous soils of Baiyin region, Gansu province, Pr China. *Water, Air, and Soil Pollution, 118*, 131–142.

Nguyen, T. A., Liu, Y., Wu, S., & Huang, L. (2022). Unravelling in-situ hardpan properties and functions in capping sulfidic cu-Pb-Zn tailings and forming a duplex soil system cover. *Journal of Hazardous Materials, 425*, 127943.

Nie, M., Xian, N., Fu, X., Chen, X., & Li, B. (2010). The interactive effects of petroleum-hydrocarbon spillage and plant rhizosphere on concentrations and distribution of heavy metals in sediments in the Yellow River Delta, China. *Journal of Hazardous Materials, 174*(1–3), 156–161.

Peng, S. P., & Bi, Y. L. (2020). Strategic consideration and core technology about environmental ecological restoration in coal mine areas in the Yellow River basin of China. *Journal of China Coal Society, 45*(4), 1. (In Chinese with English abstract).

Pilon-Smits, E. (2005). Phytoremediation. *Annual Review of Plant Biology, 56*(1), 15–39.

Qiu, J., Zhu, M., Zhou, Y., & Guan, X. (2021). Effect and mechanism of coal gangue concrete modification by fly ash. *Construction and Building Materials, 294*, 123563.

Salt, D. E., Smith, R. D., & Raskin, I. (1998). Phytoremediation. *Annual Review of Plant Biology, 49*(1), 643–668.

Su, H., Kang, W., Xu, Y., & Wang, J. (2018). Assessing groundwater quality and health risks of nitrogen pollution in the Shenfu mining area of Shaanxi Province, Northwest China. *Exposure and Health, 10*, 77–97.

Tan, W. J., Zhao, G. B., Wei, J. S., Wang, X. H., Jiang, H. B., & Peng, S. X. (2023). Characteristics, distribution and utilization potential of mineral resources in the Yellow River Basin. *Northwestern Geology, 56*(2), 163–174. (in Chinese).

Wang, C., Wang, W., He, S., Du, J., & Sun, Z. (2011). Sources and distribution of aliphatic and polycyclic aromatic hydrocarbons in Yellow River Delta nature reserve, China. *Applied Geochemistry, 26*(8), 1330–1336.

Wang, M., Qi, S., & Zhang, X. (2012). Wetland loss and degradation in the Yellow River delta, Shandong Province of China. *Environmental Earth Sciences, 67*, 185–188.

Wang, S., Nan, Z., Prete, D., Ma, J., Liao, Q., & Zhang, Q. (2016). Accumulation, transfer, and potential sources of mercury in the soil-wheat system under field conditions over the loess plateau, Northwest China. *Science of the Total Environment, 568*, 245–252.

Wang, Y., Liang, J., Wang, J., & Gao, S. (2018). Combining stable carbon isotope analysis and petroleum-fingerprinting to evaluate petroleum contamination in the Yanchang oilfield located on loess plateau in China. *Environmental Science and Pollution Research, 25*, 2830–2841.

Wang, S., Liu, T., Xiao, X., & Luo, S. (2021). Advances in microbial remediation for heavy metal treatment: A mini review. *Journal of Leather Science and Engineering, 3*, 1–10.

Wang, N., Luo, Y., Liu, Z., & Sun, Y. (2022). Spatial distribution characteristics and evaluation of soil pollution in coal mine areas in loess plateau of northern Shaanxi. *Scientific Reports, 12*(1), 16440.

Wohlfart, C., Mack, B., Liu, G., & Kuenzer, C. (2017). Multi-faceted land cover and land use change analyses in the Yellow River Basin based on dense Landsat time series: Exemplary analysis in mining, agriculture, forest, and urban areas. *Applied Geography, 85*, 73–88.

Wu, Q., Zhou, W., Wang, J., & Xie, S. (2009). Prediction of groundwater inrush into coal mines from aquifers underlying the coal seams in China: Application of vulnerability index method to Zhangcun coal mine, China. *Environmental Geology, 57*, 1187–1195.

Wu, Q., Liu, Y., Liu, D., & Zhou, W. (2011). Prediction of floor water inrush: The application of GIS-based AHP vulnerable index method to Donghuantuo coal mine, China. *Rock Mechanics and Rock Engineering, 44*, 591–600.

Wu, H., Wen, Q., Hu, L., Gong, M., & Tang, Z. (2017). Feasibility study on the application of coal gangue as landfill liner material. *Waste Management, 63*, 161–171.

Wu, J., Zhou, H., He, S., & Zhang, Y. (2019). Comprehensive understanding of groundwater quality for domestic and agricultural purposes in terms of health risks in a coal mine area of the Ordos basin, north of the Chinese loess plateau. *Environmental Earth Sciences, 78*, 1–17.

Xu, H., Xie, W. J., & Lu, Z. H. (2013). Petroleum contaminated soil remediation using six wild plant species in the Yellow River Delta. *Applied Mechanics and Materials, 246*, 598–601.

Yang, Z., Li, W., Li, X., Wang, Q., & He, J. (2019). Assessment of eco-geo-environment quality using multivariate data: A case study in a coal mining area of Western China. *Ecological Indicators, 107*, 105651.

Yang, D., Qiu, H., Ma, S., Liu, Z., Du, C., Zhu, Y., & Cao, M. (2022). Slow surface subsidence and its impact on shallow loess landslides in a coal mining area. *Catena, 209*, 105830.

Yu, S., Li, S., Tang, Y., & Wu, X. (2011). Succession of bacterial community along with the removal of heavy crude oil pollutants by multiple biostimulation treatments in the Yellow River Delta, China. *Journal of Environmental Sciences, 23*(9), 1533–1543.

Yuan, Y., Zhao, Z., Niu, S., Li, X., Wang, Y., & Bai, Z. (2018). Reclamation promotes the succession of the soil and vegetation in opencast coal mine: A case study from Robinia pseudoacacia reclaimed forests, Pingshuo mine, China. *Catena, 165*, 72–79.

Yuan, L., Gao, Y., Cheng, F., Du, J., Hu, Z., Yang, X., Wang, H., Kong, X., Fu, Z., Wang, W., & Duan, L. (2022). The influence of oil exploitation on the degradation of vegetation: A case study in the Yellow River Delta nature reserve, China. *Environmental Technology & Innovation, 28*, 102579.

Zang, F., Wang, S., Nan, Z., Ma, J., Zhang, Q., Chen, Y., & Li, Y. (2017). Accumulation, spatio-temporal distribution, and risk assessment of heavy metals in the soil-corn system around a polymetallic mining area from the loess plateau, Northwest China. *Geoderma, 305*, 188–196.

Zhai, X., Wu, S., Wang, K., Drebenstedt, C., & Zhao, J. (2017). Environment influences and extinguish technology of spontaneous combustion of coal gangue heap of Baijigou coal mine in China. *Energy Procedia, 136*, 66–72.

Zhang, Y., & Ling, T. C. (2020). Reactivity activation of waste coal gangue and its impact on the properties of cement-based materials–a review. *Construction and Building Materials, 234*, 117424.

Zhang, X., She, D., Wang, G., & Huang, X. (2021). Source identification of soil elements and risk assessment of trace elements under different land uses on the loess plateau, China. *Environmental Geochemistry and Health, 43*, 2377–2392.

Zhao, G., Wu, T., Ren, G., Zhu, Z., Gao, Y., Shi, M., Ding, S., & Fan, H. (2023). Reusing waste coal gangue to improve the dispersivity and mechanical properties of dispersive soil. *Journal of Cleaner Production, 404*, 136993.

Zhen, Q., Zheng, J., & Zhang, X. (2019). Changes of solute transport characteristics in soil profile after mining at an opencast coal mine site on the loess plateau, China. *Science of the Total Environment, 665*, 142–152.

Zhou, B. B., Shao, M., Wen, M. X., Wang, Q. J., & Horton, R. (2010). Effects of coal gangue content on water movement and solute transport in a China loess plateau soil. *CLEAN–Soil, Air Water, 38*(11), 1031–1038.

Chapter 12
Policy-driven Large Scale Environmental Restoration Projects

Contents

12.1 Introduction

Understanding environmental problems and their underlying causes is a fundamental prerequisite for effective control and remediation. The implementation of practical solutions relies heavily on robust social, economic, and environmental policies. The primary goal of watershed management is to achieve sustainable development by enhancing people's well-being, promoting societal and economic growth, and ensuring ecological restoration. However, the complexity and severity of environmental issues in the Yellow River basin presents significant challenges in achieving this goal. These challenges are further exacerbated by the region's relatively low economic development compared to other parts of the country, making it even more difficult to achieve the previously mentioned goal.

This chapter introduces several large-scale ecological restoration projects implemented in the Yellow River basin. It explores the related legislative policies, evaluates their social, economic, and environmental impacts, and investigates stakeholder involvement in watershed management. The chapter also examines the collaboration between the government and farmers, the balance between short-term and

T. Zhu, *Watershed Environmental Degradation and Sustainable Development*,
https://doi.org/10.1007/978-3-032-02448-0_12

long-term development goals, and the social and economic benefits and conflicts arising from watershed management in the region.

12.2 Grain for Green Programme

The Grain for Green Program (GFGP), launched by the Chinese government in 1999, aims to combat soil erosion, land degradation, and vegetation loss while also addressing rural poverty, particularly in ecologically fragile areas. The Yellow River basin, especially the Loess Plateau in its upper and middle reaches, is one of the most severely eroded regions in China and was prioritized for this project. Rapid population growth in the area led to the widespread clearance of natural vegetation for agriculture, including on steep slopes. However, these steep-slope croplands are characterized by low crop productivity and high erosion rates, making the Loess Plateau a primary region for GFGP implementation. The government offers monetary and food subsidies to farmers who convert sloped croplands into vegetated lands. These subsidies include 1,500 kg of grain per hectare of converted farmland per year, a living allowance of 300 yuan per hectare per year, and a one-time grant of 750 yuan per hectare for seeds or seedlings. The subsidy duration varies based on the type of conversion: 2 years for grassland, 5 years for commercial forests, and 8 years for ecological forests (Liu et al., 2008; Wu et al., 2019).

The Grain for Green Program (GFGP) has significantly transformed land use in the Loess Plateau, converting vast areas of sloped cropland and degraded land into vegetated landscapes. Numerous studies have attempted to quantify these changes, yielding varying results. Deng et al. (2019) reported that since 2000, 8.6 million hectares of degraded land have been transformed into forests, shrubs, and grasslands under the GFGP. Fu et al. (2017) noted a 12.94% reduction in sloped cropland area between 2000 and 2010, while Feng et al. (2016) observed a 25% increase in vegetative cover, including forests and grasses. Using satellite imagery, Xiao (2014) found that tree cover on the plateau rose from 7.44% to 10.49% during this period, reflecting a 41.0% increase. The most significant afforestation took place in the southern, southeastern, and eastern parts of the region, while the northern and northwestern areas saw minimal changes. This spatial variation aligns with the precipitation gradient across the Loess Plateau, with drier areas in the northwest being less suitable for afforestation.

The Grain for Green Program (GFGP) has played an important role in reducing soil erosion in the Loess Plateau region. By increasing vegetation cover, soil erodibility has significantly decreased. Between 2001 and 2018, the average annual sediment load of the Yellow River dropped by 85% compared to levels recorded from 1919 to 1959 (Zhao et al., 2021). This reduction may be attributed to changes in precipitation patterns and the implementation of various soil and water conservation measures, such as revegetation, check dams, and terraces. Wang et al. (2022) estimated that human activities accounted for 93% of the sediment reduction over the past two decades. Fu et al. (2017) calculated that vegetation restoration under the

GFGP reduced soil loss by 0.15 Gt per year in the Loess Plateau. While revegetation is highly effective in mitigating slope water erosion, it may lead to an increase in gravity-induced erosion, such as shallow landslides (Guo et al., 2020).

The conversion of sloped cropland into vegetation-covered land can greatly enhance soil organic carbon (SOC) storage in the region. However, the type of vegetation significantly influences soil carbon sequestration due to variations in ecological processes and succession pathways. In forested areas, aboveground biomass serves as the primary contributor to soil carbon, while tree roots have a minimal impact. In contrast, grass roots are the main source of soil carbon in grasslands. Additionally, nutrient recycling occurs more rapidly in grasses compared to trees. Consequently, the topsoil in grasslands generally contains higher levels of soil carbon than that in forested areas. Wei et al. (2012) compared soil organic carbon (SOC) levels between restored forestland and grassland in the Loess Plateau region. Their findings revealed that SOC in the top 20 cm of grassland soil was more than double that of forestland soil. This suggests that naturally restored grasslands may be a more effective vegetation type for SOC sequestration, primarily due to higher carbon input from roots. Compared to sloped cropland, carbon loss from vegetated lands is significantly lower due to reduced sediment and runoff-related carbon losses (Deng et al., 2014). Deng et al. reported that SOC loss is primarily driven by sediment transfer (99.6%), with runoff accounting for only a minor portion (0.4%). As a result, under the Grain for Green Program (GFGP) in the Loess Plateau, SOC in the top 20 cm of soil is projected to increase at a rate of 0.712 TgC per year over a 60-year period (Chang et al., 2011).

The Grain for Green Program (GFGP) has had significant economic and social impacts on local farmers. The retirement of steep slopeland and the prohibition of grazing on vulnerable grasslands under the program have directly affected the livelihoods of rural residents, leading to mixed outcomes. A case study by Shi and Wang (2011) found that 69% of farm households in a northern county of Shaanxi province increased their net incomes by 2–5 times over a 10-year period following the program's implementation. However, over half of these households' income came from family members working as migrant workers. This reliance poses a risk, as diminishing urban job opportunities could drive these farmers back into poverty. Wu et al. (2021) criticized the program for prioritizing ecological goals over economic ones, arguing that it has reduced rural household incomes in some regions. This decline is attributed to significant losses in grazing income, inadequate compensation, the failure of economic forestry initiatives, and ineffective labor training programs.

Food security must be carefully evaluated in the context of the Grain for Green Program (GFGP). Following the program's implementation, the total cropland area in the Loess Plateau decreased by 17.3% between 1998 and 2003 (Lyu & Xu, 2020). Grain production in the region is influenced by various factors, including total cropland area and grain yield per unit area. The latter depends on several factors such as soil fertility, fertilizer and pesticide use, and irrigation, among others. To offset the reduction in cropland, grain yield per unit area must increase to sustain the region's large population (Shi et al., 2020; Geng et al., 2020). Check-dam land and terrace land are more productive than slope cropland due to higher soil nutrient content, better moisture retention, and greater resistance to droughts, which are common in this semi-arid region. Studies on the impact of the GFGP on total grain production in the Loess Plateau have produced mixed results, ranging from significant or modest reductions to slight increases (Feng et al., 2005; Xu et al., 2006; Deng & Shuangguan, 2011; Lyu & Xu, 2020).

The afforestation efforts under the Grain for Green Program (GFGP) have raised concerns about water availability (Jia et al., 2017; Zhang et al., 2018a, 2018b). Converting cropland into shrubland or woodland can negatively impact soil moisture due to increased evapotranspiration. Soil moisture data collected from 50 sites across the Loess Plateau revealed a significant decline in soil water storage, particularly over a 4-meter soil profile, after the introduction of *Caragana korshinskii*, one of the most commonly planted exotic shrub species in the region, compared to cropland (Jia et al., 2017). Tree plantations can lead to severe dryness in both shallow and deep soil layers (Jia & Shao, 2014). In regions with prolonged rainfall shortages, artificial forests can deplete deep soil water, resulting in soil desiccation and the formation of dry soil layers at depth (Jiang et al., 2019). Therefore, careful consideration must be given to the location and type of vegetation selected for revegetation projects. Generally, artificial forests are only suitable for wetter areas of the Loess Plateau, where annual precipitation exceeds 500 mm. It is advisable to prioritize tree species with lower water consumption to ensure that forest water use does not surpass the region's available water resources. In areas with lower precipitation, natural grassland revegetation is a more sustainable option (Ren et al., 2018).

Revegetation alters local water cycles by increasing infiltration, evaporation, and transpiration, which in turn reduces runoff and soil moisture. However, the impact of large-scale revegetation under the Grain for Green Program (GFGP) on regional climate remains a topic of debate. A recent study suggests that precipitation increased by 54.62 mm following the initiation of large-scale afforestation since 1999, potentially due to enhanced atmospheric moisture from transpiration (Tian et al., 2022). Nevertheless, it is challenging to determine whether this precipitation increase is primarily attributable to afforestation efforts or natural rainfall fluctuations.

12.3 Three-river Headwaters Nature Reserve

The Three-River Headwaters Region (TRHR), the largest nature reserve in China, is located in the heart of the Qinghai-Tibet Plateau. Known as the "Chinese Water Tower," it serves as the source of the Yellow, Yangtze, and Mekong Rivers and contributes 49% of the Yellow River's water supply (Zhang et al., 2012). The region's terrain consists of high mountains, deep gorges, and flat plateaus, with an average elevation of about 4,400 m, ranging from 3,335 m to 6,564 m. The climate is characterized as a continental alpine type, with cold, dry winters and wet, mild summers. Annual average temperatures range from −5.6 °C to 3.8 °C, with the warmest month being July (6.4 °C to 13.2 °C) and the coldest month January (−6.6 °C to

−13.8 °C). Annual precipitation varies from 262.2 mm to 772.8 mm, with 75% falling between June and September. Annual potential evaporation ranges from 730 mm to 1,700 mm (Three-River Source Region Nature Reserve, 2008). This area is a critical hotspot for alpine biodiversity, home to 69 nationally protected species, including 17 at level-I protection and 52 at level-II protection (Shao et al., 2013; Liu et al., 2014; Jiang & Zhang, 2016a, 2016b). The landscape is predominantly grassland and meadow, containing the world's largest alpine wetland ecosystem. The Yellow River source region covers a total area of 137,700 km^2, encompassing 15 counties in Qinghai Province.

The Three-River Headwaters Region (TRHR) faces numerous environmental challenges due to climate change and human activities, including glacial melting, wetland destruction, grassland and permafrost degradation, soil erosion, and desertification (Li et al., 2013; Li et al., 2014; Li et al., 2016). To address these issues, the Chinese government launched the "Overall Planning of Eco-environment Protection and Construction in the TRHR" project in 2005, with a total investment of $11.8 billion USD (Jiang & Zhang, 2016a, 2016b). The project established a nature reserve covering 152,000 km^2. The TRHR program focuses on protecting and restoring degraded ecosystems by reducing livestock numbers, particularly in alpine meadow areas, and increasing grassland coverage to relieve overgrazing pressures. The reserve is divided into core, buffer, and experimental areas to balance conservation and sustainable use.

- Core Areas (20.5%): Strictly protected zones to conserve critical ecosystems.
- Buffer Areas (25.8%): Surround the core zones to minimize the impact of nearby towns and industrial areas while serving as wildlife corridors.
- Experimental Areas (53.7%): Located outside the core and buffer zones, these areas aim to combine ecosystem protection with sustainable land use, supporting the improvement of local livelihoods (Ran et al., 2016).

Numerous studies have evaluated the impact of the project on vegetation recovery, but their findings remain contentious. Additionally, distinguishing the relative contributions of natural and anthropogenic factors to these changes is challenging. Zhang et al. (2017) reported an 11.4% increase in vegetation cover and a 30.3% rise in mean annual grass yield following the project's implementation. Similarly, Jiang and Zhang (2016b) observed a significant improvement in carbon sequestration attributed to increased Net Primary Production (NPP), with a notably greater rise in NPP within the reserve area compared to non-reserve areas, where NPP also showed some improvement. However, Shen et al. (2018) indicated that while the project had a positive effect on vegetation recovery, this impact was statistically insignificant when comparing vegetation changes inside and outside the reserve from 2005 to 2015.

The impact of the project on desertification has been most evident in the drier parts of the region, particularly in the source area of the Yellow River. Between 1990 and 2005, desertification remained relatively stable in the central part of the Yellow River source region but increased in the eastern and southern parts. After 2005, desertification in the upper reaches of the Yellow River source region began to improve gradually, with significant progress observed in the northwestern parts (Guo et al., 2022). Sun et al. (2019) studied the process of grassland desertification in Northern Tibet, including the Yellow River source area, and found that while grassland conditions improved overall during 2001–2015, the extent of improvement was smaller than the area designated for restoration. Similarly, Ren et al. (2016) reported that desertification vulnerability in the Yellow River source region increased between 2000 and 2005 but declined from 2005 to 2010. They attributed the initial rise in vulnerability primarily to rising temperatures, while the ecological restoration project, implemented in 2005, played a significant role in controlling desertification thereafter.

The impact of the TRHR project on streamflow is complex and multifaceted. Some studies indicate a steady increase in streamflow generated from the TRHR. Following the implementation of the project in 2005, the annual mean streamflow (2005–2012) in the Yangtze, Yellow, and Mekong Rivers was 39.3%, 5.2%, and 9.7% higher, respectively, than the mean levels recorded between 1956 and 2012 (Jiang & Zhang, 2016a, 2016b). This increase reflects the combined effects of climate change and human activities. Increased precipitation is a primary factor contributing to higher streamflow discharge. Rising temperatures have also intensified glaciers melting and snowmelt, further boosting water flow. Additionally, warmer temperatures have deepened the active layer in permafrost regions, increasing groundwater storage through infiltration. Urbanization has contributed to higher surface runoff as well. Conversely, other factors act to reduce water availability. Rising temperatures enhance evaporation, and the conversion of farmland to grassland reduces surface runoff. Moreover, some studies suggest a decline in overall water availability during the same period, driven by reduced precipitation during the wet season and increasing temperatures (Hu et al., 2011; Meng et al., 2016).

To enhance residents' livelihoods and well-being while safeguarding ecosystems, the Chinese State Council approved the "State Integrated Test and Demonstration Region (SITDR) of the TRHR" in 2011. This initiative encompasses four Tibetan Autonomous Prefectures and 21 counties in Qinghai Province, covering a total area of 394,650 km^2, equivalent to 54.5% of the province's total area (Fang, 2013). To alleviate grazing pressure on grasslands, the Chinese government set a goal to relocate 100,000 herdsmen out of the TRHR protection zones. Each household was provided with subsidies, free housing, and job opportunities. By the end of 2007, 61,889 individuals and 13,305 households had been resettled into cities and towns, supported by the establishment of 86 migration communities. This represents the largest ecological migration effort in China (Wang et al., 2010). Between 2004 and 2012, the average household net income in the TRHR increased by 10% (Ran et al., 2016).

A key objective of the TRHR project is to protect biodiversity (Li et al., 2022). Biodiversity conservation, ecosystem services, and sustainable development are intricately connected. Conservation efforts should prioritize areas that provide the greatest benefits for habitat protection, species preservation, and ecosystem services, while promoting sustainable development (Zhang et al., 2015). Currently, the region encompasses 25 core conservation areas, including 8 wetland zones, 10 forest zones, 2 alpine meadow zones, and 5 wildlife zones.

12.4 Yellow River Delta National Natural Reserve

The Yellow River delta, one of the world's youngest deltas, is located where the Yellow River flows into the Bohai Sea, covering an area of approximately 5,400 km^2. The modern delta began forming in 1855 when flooding caused the river to change course from the Yellow Sea to the Bohai Sea. In 1976, the government artificially redirected the river's course from the north to the east. This sediment deposition led to the deltaic lobe expanding at an average rate of 1,600 m per year, with an annual land increase of 20–25 km^2 until 1996 (Saito et al., 2000). After 1996, sediment deposition decreased significantly due to a sharp reduction in sediment load from the Yellow River, which declined from 898 million tons per year in 1976 to just 31 million tons per year by 2015 (Zhang et al., 2018a, 2018b). This reduction was attributed to extensive conservation measures implemented in the river's middle reaches and a decline in precipitation. Consequently, the total area of the delta stabilized as wave erosion and sediment deposition balanced each other (Ji et al., 1994; Fang & Xu, 2000). However, future sea-level rise poses a threat to the delta's land area, potentially leading to further reductions.

The Yellow River Delta has undergone significant land use changes in recent years, driven by rapid population growth, excessive land reclamation, and land conversion. Before 1991, when the Yellow River Delta National Nature Reserve was established, the region primarily consisted of grassland, tidal flats, and unirrigated farmland, with agricultural land accounting for 47.5% of the total area (Fang & Xu, 2000). Dongying City, the largest urban center in the delta, saw its population grow from 1.793 million in 2000 to 2.209 million by 2020. Population growth and economic activities have led to the conversion of wetlands into cropland, pastureland, shrimp farms, and other uses. Cropland expanded from 467.6 km^2 in 1986 to 718.5 km^2 in 2015, with an average annual growth rate of 8.65 km^2 (Feng et al., 2015). Between 1995 and 2010, aquaculture farming and urban sprawl increased significantly, adding 1,803.1 km^2 and 237.0 km^2 of land use, respectively, to the delta (Ottinger et al., 2013). The delta is also home to China's second-largest oil field, Shengli Oilfield, which has disrupted vegetation distribution and contributed to degradation (Yuan et al., 2022). Human disturbances have intensified environmental challenges in the delta, including wetland degradation, soil and water contamination, land salinization, seawater intrusion, and biodiversity loss. Wetlands have become increasingly fragmented due to the construction of artificial dykes. The loss of natural wetlands has outpaced the creation of human-made wetlands (Liu et al., 2013; Fu et al., 2013). These changes highlight the conflict interests between development and ecological preservation in the region.

To address the environmental challenges mentioned above, the Yellow River Delta National Nature Reserve was established in 1992 to protect the newly developed wetland ecosystem in the delta region. The reserve spans a total area of 153,000 hectares, divided into three zones: the Core Area (79,000 ha), the Buffer Area (11,000 ha), and the Experimental Area (82,000 ha) (Wang et al., 2011). The reserve is split into two sections: the southern part, covering 100,450 hectares at the mouth of the current Yellow River flowing into the Bohai Sea, and the northern part, covering 48,500 hectares at the mouth of the old Yellow River channel, which changed course in 1976. Satellite imagery reveals that grassland area decreased by 50% after the reserve's establishment in the early 1990s due to its conversion into wetlands (Fang & Xu, 2000). However, the wetland area within the reserve has expanded by 12.3%, or 188 km^2, since 2017 (Xinhua News, 2021). The Yellow River Delta is critical for biodiversity conservation and serves as a vital resting and breeding ground for migratory birds along the West Pacific Flyway. The reserve is home to numerous protected species, including seven first-class national protection bird species such as *Grus japonensis*, *Otis tarda*, *Haliaeetus albicilla*, *Ciconia boyciana*, and *Mergus squamatus*. For instance, the endangered white stork (*C. boyciana*), listed by the International Union for Conservation of Nature (IUCN), uses the reserve as a breeding and resting site during seasonal migration. Since 2002, more than 2,200 white stork chicks have been born in the reserve. Overall, 1,630 species of wild animals and 685 species of plants have been recorded in the reserve. Substantial restoration efforts have been carried out, including releasing freshwater from reservoirs to recharge groundwater, converting farmland back into wetlands, and control salinization, among others. These efforts have not only improved the ecosystem but also boosted tourism, generating economic benefits for the region.

12.5 Recommendations

These large policy-driven initiatives play a vital role in restoring the fragile ecosystems of the Yellow River basin. Programs such as the Three-River Headwater Source Region Protection Program and the Yellow River Delta Nature Reserve Program focus on preserving and protecting natural habitats in alpine meadows and coastal wetlands. Coordination among these programs is equally essential to maximize their combined benefits. Protecting wetlands and restoring degraded grasslands in the Yellow River's headwater region helps minimize surface erosion, ensuring cleaner water flows downstream. Likewise, the Grain for Green Program effectively reduces soil erosion in the river's primary sediment source areas, decreasing siltation in the lower reaches of the Yellow River channel and mitigating flood hazards.

The implementation of these programs has significant implications for balancing short-term economic benefits, healthy ecosystems, and long-term sustainability. While these initiatives are government-led and funded, active and long-term participation from all stakeholders is essential to ensure that the economic and environmental benefits are sustained. To achieve sustainability, improving the livelihoods of local communities must be prioritized during project implementation. In the short term, this can be supported through government subsidies, but in the long term, self-reliance must be fostered. Furthermore, short-term and long-term restoration strategies should be carefully coordinated to maximize their effectiveness and ensure enduring success.

Extensive research has been conducted to evaluate the benefits of these programs, but findings often vary significantly and are sometimes contradictory. These discrepancies arise from differences in data sources, methodologies, and the indicators used. To address this, more long-term experimental stations should be established to monitor program effectiveness. Research and conservation efforts need to be more closely integrated into project design and planning. Additionally, spatial variations in environmental conditions should be carefully considered when developing conservation measures. Unfortunately, the top-down approach to implementation often overlooks these variations. Active involvement of residents in planning is important. Without their engagement, the long-term effectiveness of ecosystem restoration efforts may be compromised once the project ends.

References

Chang, R., Fu, B., Liu, G., & Liu, S. (2011). Soil carbon sequestration potential for "grain for green" project in loess plateau, China. *Environmental Management, 48*, 1158–1172.

Deng, L., & Shuangguan, Z. P. (2011). Food security and farmers' income: Impacts of the grain for green programme on rural households in China. *Journal of Food, Agriculture & Environment, 9*(3–4), 826–831.

Deng, L., Shangguan, Z. P., & Sweeney, S. (2014). "Grain for green" driven land use change and carbon sequestration on the loess plateau, China. *Scientific Reports, 4*(1), 7039.

Deng, L., Kim, D. G., Li, M., Huang, C., Liu, Q., Cheng, M., Shangguan, Z., & Peng, C. (2019). Land-use changes driven by 'grain for Green' program reduced carbon loss induced by soil erosion on the loess plateau of China. *Global and Planetary Change, 177*, 101–115.

Fang, Y. (2013). Managing the three-rivers headwater region, China: From ecological engineering to social engineering. *Ambio, 42*(5), 566–576.

Fang, H., & Xu, J. (2000). Land cover and vegetation change in the Yellow River Delta nature reserve analyzed with Landsat thematic mapper data. *Geocarto International, 15*(4), 43–50.

Feng, Z. M., Yang, Y. Z., Zhang, Y. Q., et al. (2005). Grain-for-green policy and its impacts on grain supply in West China. *Land Use Policy, 22*(4), 301–312.

Feng, Q., Gong, J., Liu, J., & Li, Y. (2015). Monitoring cropland dynamics of the Yellow River Delta based on multi-temporal Landsat imagery over 1986 to 2015. *Sustainability, 7*(11), 14834–14858.

Feng, X., Fu, B., Piao, S., Wang, S., Ciais, P., Zeng, Z., Lü, Y., Zeng, Y., Li, Y., Jiang, X., & Wu, B. (2016). Revegetation in China's loess plateau is approaching sustainable water resource limits. *Nature Climate Change, 6*(11), 1019–1022.

Fu, X., Liu, G. H., Chai, S. Y., et al. (2013). Spatial-temporal analysis of wetland landscape pattern under the influence of artificial dykes in the Yellow River delta. *Chinese Journal of Population Resources and Environment, 11*(2), 109–117.

Fu, B., Wang, S., Liu, Y., Liu, J., Liang, W., & Miao, C. (2017). Hydrogeomorphic ecosystem responses to natural and anthropogenic changes in the loess plateau of China. *Annual Review of Earth and Planetary Sciences, 45*, 223–243.

Geng, Q., Ren, Q., Yan, H., Li, L., Zhao, X., Mu, X., Wu, P., & Yu, Q. (2020). Target areas for harmonizing the grain for green programme in China's loess plateau. *Land Degradation & Development, 31*(3), 325–333.

Guo, W. Z., Chen, Z. X., Wang, W. L., Gao, W. W., Guo, M. M., Kang, H. L., Li, P. F., Wang, W. X., & Zhao, M. (2020). Telling a different story: The promote role of vegetation in the initiation of shallow landslides during rainfall on the Chinese loess plateau. *Geomorphology, 350*, 106879.

Guo, B., Wei, C., Yu, Y., Liu, Y., Li, J., Meng, C., & Cai, Y. (2022). The dominant influencing factors of desertification changes in the source region of Yellow River: Climate change or human activity? *Science of the Total Environment, 813*, 152512.

Hu, Y., Maskey, S., Uhlenbrook, S., & Zhao, H. (2011). Streamflow trends and climate linkages in the source region of the Yellow River, China. *Hydrological Processes, 25*(22), 3399–3411.

Ji, Z., Hu, C., Zeng, Q., & Yan, Y. (1994). Study the recent Yellow River mouth evolution using remote sensing images. *Sand Research (in Chinese), 1994*(3), 12–22.

Jia, Y. H., & Shao, M. A. (2014). Dynamics of deep soil moisture in response to vegetational restoration on the loess plateau of China. *Journal of Hydrology, 519*, 523–531.

Jia, X., Zhu, Y., & Luo, Y. (2017). Soil moisture decline due to afforestation across the loess plateau, China. *Journal of Hydrology, 546*, 113–122.

Jiang, C., & Zhang, L. (2016a). Ecosystem change assessment in the three-river headwater region, China: Patterns, causes, and implications. *Ecological Engineering, 93*, 24–36.

Jiang, C., & Zhang, L. (2016b). Effect of ecological restoration and climate change on ecosystems: A case study in the three-Rivers headwater region, China. *Environmental Monitoring and Assessment, 188*, 1–20.

Jiang, C., Zhang, H., Wang, X., Feng, Y., & Labzovskii, L. (2019). Challenging the land degradation in China's loess plateau: Benefits, limitations, sustainability, and adaptive strategies of soil and water conservation. *Ecological Engineering, 127*, 135–150.

Li, X. L., Gao, J., Brierley, G., Qiao, Y. M., Zhang, J., & Yang, Y. W. (2013). Rangeland degradation on the Qinghai-Tibet plateau: Implications for rehabilitation. *Land Degradation & Development, 24*(1), 72–80.

Li, X. L., Perry, G. L. W., Brierley, G., Sun, H. Q., Li, C. H., & Lu, G. X. (2014). Quantitative assessment of degradation classifications for degraded alpine meadows (heitutan), Sanjiangyuan, western China. *Land Degradation & Development, 25*(5), 417–427.

Li, X., Perry, G., & Brierley, G. J. (2016). Grassland ecosystems of the Yellow River Source Zone: degradation and restoration. In *Landscape and ecosystem diversity, dynamics and management in the Yellow River Source Zone* (pp. 137–165). Springer.

Li, S., Yu, D., Huang, T., & Hao, R. (2022). Identifying priority conservation areas based on comprehensive consideration of biodivers and ecosystem services in the Three-River headwaters region, China. *Journal of Cleaner Production, 359*, 132082.

Liu, J., Li, S., Ouyang, Z., Tam, C., & Chen, X. (2008). Ecological and socioeconomic effects of China's policies for ecosystem services. *Proceedings of the National Academy of Sciences, 105*(28), 9477–9482.

Liu, Y. S., Wang, G. D., & Zhang, F. G. (2013). Spatial-temporal dynamic patterns of rural area development in eastern coastal China. *Chinese Geographical Science, 23*(2), 173–181.

Liu, X., Zhang, J., Zhu, X., Pan, Y., Liu, Y., Zhang, D., & Lin, Z. (2014). Spatiotemporal changes in vegetation coverage and its driving factors in the Three-River headwaters region during 2000–2011. *Journal of Geographical Sciences, 24*, 288–302.

Lyu, C., & Xu, Z. (2020). Crop production changes and the impact of grain for green program in the loess plateau of China. *Journal of Arid Land, 12*, 18–28.

Meng, F., Su, F., Yang, D., Tong, K., & Hao, Z. (2016). Impacts of recent climate change on the hydrology in the source region of the Yellow River basin. *Journal of Hydrology: Regional Studies, 6*, 66–81.

Ottinger, M., Kuenzer, C., Liu, G., Wang, S., & Dech, S. (2013). Monitoring land cover dynamics in the Yellow River Delta from 1995 to 2010 based on Landsat 5 TM. *Applied Geography, 44*, 53–68.

Ran, Y., Xie, J., & Li, X. (2016). Socio-economic development and its effects on the ecological environment of the Yellow River Source Zone. In *Landscape and ecosystem diversity, dynamics and management in the Yellow River Source Zone* (pp. 331–351). Springer.

Ren, X., Dong, Z., Hu, G., Zhang, D., & Li, Q. (2016). A GIS-based assessment of vulnerability to aeolian desertification in the source areas of the Yangtze and yellow Rivers. *Remote Sensing, 8*(8), 626.

Ren, Z., Li, Z., Liu, X., Li, P., Cheng, S., & Xu, G. (2018). Comparing watershed afforestation and natural revegetation impacts on soil moisture in the semiarid loess plateau of China. *Scientific Reports, 8*(1), 2972.

Saito, Y., Wei, H., Zhou, Y., Nishimura, A., Sato, Y., & Yokota, S. (2000). Delta progradation and chenier formation in the Huanghe (Yellow River) delta, China. *Journal of Asian Earth Sciences, 18*(4), 489–497.

Shao, Q. Q., Liu, J. Y., Huang, L., Fan, J. W., Xu, X. L., & Wang, J. B. (2013). Integrated assessment on the effectiveness of ecological conservation in Sanjiangyuan National Nature Reserve. *Geographical Research, 32*(9), 1645–1656.

Shen, X., An, R., Feng, L., Ye, N., Zhu, L., & Li, M. (2018). Vegetation changes in the three-river headwaters region of the Tibetan plateau of China. *Ecological Indicators, 93*, 804–812.

Shi, W., & Wang, K. (2011). Assessment of ecological, economic and social impacts of grain for green on the counties of North Shaanxi in the loess plateau, China: A case study of Mizhi County. *African Journal of Biotechnology, 10*(70), 15763–15769.

Shi, P., Feng, Z., Gao, H., Li, P., Zhang, X., Zhu, T., Li, Z., Xu, G., Ren, Z., & Xiao, L. (2020). Has "grain for green" threaten food security on the loess plateau of China? *Ecosystem Health and Sustainability, 6*(1), 1709560.

Sun, J., Hou, G., Liu, M., Fu, G., Zhan, T., Zhou, H., Tsunekawa, A., & Haregeweyn, N. (2019). Effects of climatic and grazing changes on desertification of alpine grasslands, Northern Tibet. *Ecological Indicators, 107*, 105647.

Three-River Source Region Nature Reserve. (2008). *Qinghai Province Government.*

Tian, L., Zhang, B., Chen, S., Wang, X., Ma, X., & Pan, B. (2022). Large-scale afforestation enhances precipitation by intensifying the atmospheric water cycle over the Chinese loess plateau. *Journal of Geophysical Research: Atmospheres, 127*(16), e2022JD036738.

Wang, Z., Song, K., & Hu, L. (2010). China's largest scale ecological migration in the three-river headwater region. *Ambio, 39*, 443–446.

Wang, C., Wang, W., He, S., Du, J., & Sun, Z. (2011). Sources and distribution of aliphatic and polycyclic aromatic hydrocarbons in Yellow River Delta nature reserve, China. *Applied Geochemistry, 26*(8), 1330–1336.

Wang, J., Shi, B., Zhao, E., Yuan, Q., & Chen, X. (2022). The long-term spatial and temporal variations of sediment loads and their causes of the Yellow River Basin. *Catena, 209*, 105850.

Wei, J., Cheng, J., Li, W., & Liu, W. (2012). Comparing the effect of naturally restored forest and grassland on carbon sequestration and its vertical distribution in the Chinese loess plateau. *PLoS One, 7*(7), e40123.

Wu, X., Wang, S., Fu, B., Feng, X., & Chen, Y. (2019). Socio-ecological changes on the loess plateau of China after grain to green program. *Science of the Total Environment, 678*, 565–573.

Wu, Z., Dai, X., Li, B., & Hou, Y. (2021). Livelihood consequences of the grain for green Programme across regional and household scales: A case study in the loess plateau. *Land Use Policy, 111*, 105746.

Xiao, J. (2014). Satellite evidence for significant biophysical consequences of the "grain for green" program on the loess plateau in China. *Journal of Geophysical Research: Biogeosciences, 119*(12), 2261–2275.

Xinhua News. (2021). http://www.xinhuanet.com/english/2021-12/17/c_1310378555.htm

Xu, Z. G., Xu, J. T., Deng, X. Z., et al. (2006). Grain for green versus grain: Conflict between food security and conservation set-aside in China. *World Development, 34*(1), 130–148.

Yuan, L., Gao, Y., Cheng, F., Du, J., Hu, Z., Yang, X., Wang, H., Kong, X., Fu, Z., Wang, W., & Duan, L. (2022). The influence of oil exploitation on the degradation of vegetation: A case study in the Yellow River Delta nature reserve. *China. Environmental Technology & Innovation, 28*, 102579.

Zhang, Y., Zhang, S., Zhai, X., & Xia, J. (2012). Runoff variation and its response to climate change in the three Rivers source region. *Journal of Geographical Sciences, 22*, 781–794.
Zhang, L., Fu, B., Lü, Y., & Zeng, Y. (2015). Balancing multiple ecosystem services in conservation priority setting. *Landscape Ecology, 30*, 535–546.
Zhang, L., Fan, J., Zhou, D., & Zhang, H. (2017). Ecological protection and restoration program reduced grazing pressure in the Three-River headwaters region, China. *Rangeland Ecology & Management, 70*(5), 540–548.
Zhang, S., Yang, D., Yang, Y., Piao, S., Yang, H., Lei, H., & Fu, B. (2018a). Excessive afforestation and soil drying on China's loess plateau. *Journal of Geophysical Research: Biogeosciences, 123*(3), 923–935.
Zhang, X., Yang, Z., Zhang, Y., Ji, Y., Wang, H., Lv, K., & Lu, Z. (2018b). Spatial and temporal shoreline changes of the southern Yellow River (Huanghe) Delta in 1976–2016. *Marine Geology, 395*, 188–197.
Zhao, Y., Hu, C., Zhang, X., Lv, X., Yin, X., & Wang, Z. (2021). Response of sediment discharge to soil erosion control in the middle reaches of the Yellow River. *Catena, 203*, 105330.

Chapter 13
Climate Change Impacts And Adaptation

Contents

13.1 Introduction

Climate change stands as one of the most formidable and urgent environmental challenges of the 21st century. The Intergovernmental Panel on Climate Change (IPCC) highlights this in its report, noting that simulations from the Atmosphere–Ocean General Circulation Model (AOGCM) predict a rise in the global mean surface temperature by approximately 3 °C by the century's end, with a range of 1.3 °C to 4.5 °C compared to 1961–1990 levels, if governments do not take greater action on slashing planet-warming emissions (IPCC, 2019). The Yellow River basin is highly susceptible to the effects of climate change. This chapter examines its environmental impacts within the basin, as well as implications for social and economic development.

T. Zhu, *Watershed Environmental Degradation and Sustainable Development*, https://doi.org/10.1007/978-3-032-02448-0_13

13.2 The Source Region of the Yellow River

The source region of the Yellow River stretches from its origin on the northern slope of the Bayan Har Mountains (32°12′–35°48′ N, 95°50′–103°28′ E) to the upstream area of the Tangnaihai hydrological gauging station. Covering an area of 121,900 km^2, it constitutes 16% of the total basin area. Situated on the Qinghai-Tibet Plateau at elevations ranging from 2,633 to 6,253 meters above sea level, this region features diverse landscapes, including alpine meadows, grasslands, alpine shrubs, forests, deep gorges, and lakes. Its high-altitude location makes it an ecologically fragile zone. The annual mean air temperature ranges from −3.7 °C to −1.1 °C, while annual precipitation varies between 357.0 mm and 677.6 mm. The region contributes a mean annual flow discharge of 20.8 × 10^9 m^3, accounting for approximately one-third of the Yellow River's total discharge (Liu et al., 2023).

Observations indicate that the region has undergone significant warming since the 1950s, with an increase of approximately 0.16 °C per decade in the annual mean temperature and 0.32 °C per decade in the winter mean temperature. These rates exceed those observed in the Northern Hemisphere and at comparable latitudes (Liu & Chen, 2000). The warming is elevation-dependent. Liu and Chen (2000) demonstrated that the warming intensity increases with elevation. Yang et al. (2014) reported that the warming rate rises between 3,000 m and 4,800 m above sea level, stabilizes around 5,000 m, and decreases above 6,200 m on the Tibetan Plateau. The relatively stable or declining warming rates above 5,000 m are likely influenced by snow and glacial coverage, which enhances solar energy reflection and consumes solar energy through ice melting. Regarding precipitation changes, some studies reported no significant long-term trends over recent decades (Hu et al., 2012). However, other research revealed substantial decadal fluctuations, with wetter conditions observed in the 1980s and drier conditions in the 1990s (Zhou & Huang, 2012). Spatial variations in precipitation were also noted, with increases in drier areas and decreases in wetter areas compared to regional norms (Zhou & Huang, 2012).

Various models have been utilized to predict climate change in the source region of the Yellow River Basin (YRB). While climate models consistently project an increase in temperature, they differ significantly regarding precipitation trends (You et al., 2014; Lu et al., 2018). For instance, Li et al. (2012), using the PRECIS model (Providing Regional Climates for Impacts Studies), predicted a decline in future precipitation for the region. Under the A2 and B2 greenhouse gas emission scenarios—representing medium-high and medium-low emissions as defined by the IPCC in 2000—the annual mean temperature is projected to rise by 3.2 °C and 3.5 °C (A2) and 3.6 °C and 3.7 °C (B2). Precipitation is expected to decrease by 31.2% and 30.2% (A2) and 26.7% and 29.1% (B2), while evaporation is projected to decline by around 25% under both scenarios. Conversely, Sun et al. (2022), using CMIP6 models, predicted that precipitation in the YRB will increase under all emission scenarios, with the most significant increases in the western and northern regions.

The impact of climate change on water resources is critically important, as it not only influences fragile local ecosystems and socio-economic development in the source region but also triggers cascading effects downstream. Global warming alters hydrological cycles. Rising temperatures result in earlier snowmelt, increased glacier melting, and permafrost thawing, as well as heightened evaporation rates. Various hydrological models and quantitative analytical methods, integrated with field observations, have been employed to evaluate the effects of climate change on hydrological processes in the region. The models typically simulate streamflow variations and pathways over recent decades, with their accuracy validated against observed data from gauging stations. Cuo et al. (2013) found that annual streamflow, as well as streamflow during wet and warm seasons, has decreased primarily due to the impacts of climate change. Additionally, more streamflow is generated earlier in the year, which is attributed to higher summer evaporation and earlier snowmelt in spring. The extent of permafrost has shrunk, and the depth of the active layer in permafrost regions has increased, allowing greater summer infiltration. This, in turn, has reduced surface runoff while increasing baseflow. Qin et al. (2017) examined changes in frozen ground from 1981 to 2015 and reported that the areal mean maximum thickness of seasonally frozen ground ranged from 1.1 to 1.8 meters, decreasing at a rate of 1.2 cm per year. The proportion of permafrost area relative to the total area also declined by 1.1% annually. The hydrologic response varies spatially within the source region. Zhang et al. (2022) reported that at higher altitudes, total runoff tends to increase more significantly due to greater snowmelt, whereas lower-altitude areas experience increased evapotranspiration, which partially offsets potential water resource gains from increased precipitation. They estimated that rainfall flow contributed 48%–56%, snowmelt flow accounted for 26%–23%, and baseflow contributed 21%–32%.

With rising temperatures, it is anticipated that the tree lines in the alpine regions of the Tibetan Plateau will lower, allowing more tree species to grow in areas currently covered by ice and snow. However, these changes may be slow and significantly constrained by past cold events (Ming et al., 2008; Fang et al., 2009; Liang et al., 2011; Li et al., 2019). Plant species and net primary productivity in the region respond variably to the warming climate, influenced by precipitation patterns. On the one hand, higher temperatures can enhance primary productivity and extend the growing season. On the other hand, increased evaporation and limited precipitation can stress vegetation growth, leading to the degradation of grasslands and alpine meadows. Permafrost degradation further reduces soil moisture in surface layers, hindering plant growth. These challenges are compounded by socio-economic activities in the region. Studies indicate that natural grasslands have been degrading since the 1980s due to the combined impacts of global warming, population growth, grazing pressure, and rodent damage (Harris, 2010; Liu et al., 2013; Chen et al., 2014). Continuous yak grazing has been shown to reduce plant height, coverage, and species richness (Zhang et al., 2015a, 2015b). Restoring degraded grasslands requires sustainable grazing management, reseeding of degraded lands, and pest control (Dong et al., 2020). Government intervention also plays a critical role in helping local farmers adapt to climate change. This can include providing subsidies, relocating communities from ecologically fragile areas, offering training to overcome adaptation challenges, and promoting solar energy use among households to reduce CO_2 emissions (Fang & Wei, 2013; He et al., 2022; Wang et al., 2020).

13.3 The Upper and Middle Reaches

While the upper reaches traditionally refer to the area above Hekou in Inner Mongolia, in this context, they denote the region extending from the Tangnaihai hydrological station, which marks the outlet of the source region, to Hekou. The middle reaches span from Hekou to Zhengzhou in Henan Province. This region covers approximately 600,000 km^2, accounting for 79% of the total basin area. Characterized by a semi-arid and arid climate, the area is particularly vulnerable to the impacts of climate change. With a high population density and critical agricultural croplands essential for feeding the population, the region faces significant challenges. Climate change poses complex and multifaceted issues for environmental protection, agricultural productivity, and economic development, necessitating innovative and adaptive strategies.

This region, which includes the Loess Plateau, has a temperate climate ranging from semi-arid to semi-humid. The mean annual temperature ranges from 4.3 °C in the northwest to 14.3 °C in the southeast (Sun et al., 2016). Mean annual precipitation follows a similar gradient, increasing from 150 mm in the northwest to 750 mm in the southeast, with over 70% of the annual rainfall occurring during the summer. Studies indicate significant climatic changes in the region between 1961 and 2010. During this period, the mean annual temperature rose markedly by 1.91 °C over 50 years, while mean annual precipitation decreased by 29.11 mm over the same timeframe (Wang et al., 2012). The number of days with extremely warm temperatures and their duration is increasing, as is the frequency of consecutive dry days. Additionally, the growing season in the region has lengthened by 3.61 days per decade (Sun et al., 2016). Future projections suggest a steady rise in temperatures. Zhang and Liu (2005), using the Hadley Centre's general circulation model (HadCM3), predicted that by the end of this century, maximum temperatures in the southern Loess Plateau could rise by 2.3–4.3 °C, while minimum temperatures could increase by 3.6–5.3 °C. In terms of precipitation, Gao et al. (2017), using Representative Concentration Pathway (RCP) 2.6 and RCP 8.5 scenarios, projected that annual precipitation would increase by 16.4 mm to 19.1 mm per decade by 2050. Similarly, Zhang and Liu (2005) predicted a 23–37% increase in annual precipitation in the central Loess Plateau by the end of the century based on the HadCM3 model.

Climate change has significant effects on hydrology, soil erosion, vegetation, and agriculture in the region. A considerable number of studies have been conducted to evaluate the impact of past and future climate change on runoff generation and soil loss using various models. For instance, Li et al. (2009) used the SWAT model to assess the effects of land use and climate changes on an agricultural watershed in the Loess Plateau from 1981 to 2000. Their findings indicated that land use change and climate changes reduced runoff by 9.6% and 95.8%, respectively, and decreased soil water content by 18.8% and 77.1%. This suggests that climate change has a more significant impact on runoff than land use change in the watershed. During this period, the mean annual temperature increased by 6 °C, while annual precipitation decreased by 18%. Gao et al. (2016) examined the hydrological responses to climate and land use changes in the 15 main watersheds of the Loess Plateau between 1961 and 2009. Their study found that hydrological responses were generally more sensitive to land use/cover changes than to climate variability. Specifically, land use/cover was more influential in ten catchments, climate variability was the dominant factor in three catchments, and both land use/cover and climate variability had nearly equal effects in one catchment. They also pointed out that changes in precipitation have a greater impact than evapotranspiration. The differing results from the two studies are mainly due to the extent of land use changes in the specific watersheds during the study periods. In the first study, only a small portion of the watershed (4.6% of the total area) undergoing land use changes. In contrast, the second study covers a period when large-scale land use changes occurred following the initiation of the Grain for Green project. Several studies have been conducted to simulate the impact of future climate changes on soil and water loss. Li et al. (2011) used the WEPP (Water and Erosion Prediction Project) model to simulate the response of agro-ecosystems in a tableland region of the Loess Plateau. The results showed that under conventional tillage, the model predicted changes in runoff ranging from 10% to 130% and soil loss varying from −5% to 195% between 2010 and 2039. In comparison, conservation tillage was projected to change runoff by −34% to 71% and soil loss by 26% to 77% over the same period. Similarly, Zhang and Liu (2005) applied the WEPP model to predict soil and water loss in the same region by the end of the century. Their findings indicated that, under conventional tillage, runoff is expected to increase by 29% to 79%, and soil loss to rise by 2% to 81%, depending on different emission scenarios. However, adopting conservation tillage could reduce runoff by 18% to 38% and decrease soil loss by 56% to 68% compared to conventional tillage under the current climate.

Studies have been conducted to evaluate the relative effects of climate change and human activities on vegetation in the region. Kong et al. (2020) found that climate change accounted for 57.65% of the vegetation changes in the Loess Plateau between 1982 and 2015. Shi et al. (2021) reported an increasing trend in NDPI across 94.86% of the total area, with residual analysis revealing that climate change contributed 45.78% to NDVI variation, while human activities accounted for 54.22%. Li et al. (2019) examined vegetation changes in the Loess Plateau from 1982 to 2019, noting that climate change (particularly warming and drying) had negative effects on vegetation prior to the initiation of the Grain for Green Program (GGP). However, after the GGP, changes such as wetter conditions and reduced drought had positive effects on vegetation, which is consistent with an earlier study by Zhao et al. (2017). Sun et al. (2015) found that warmer temperatures promoted vegetation growth in the southern Loess Plateau, where the climate is wetter, but had a negative impact in the northern region, where conditions are drier.

Drought vulnerability is always important in semi-arid regions under climate changes. Wu et al. (2018) examined 17 catchments across the Loess Plateau and found that there was an increasing trend of drought over the period of 1961–2013. Han et al. (2021) demonstrated that the drought frequency increases from southeast to northwest over the period of 1982–2015. The droughts mainly occurred as short duration and weak intensity. Zhang et al. (2012) indicated that the areas affected by drought showed spatial difference. For the wetter part, central and southeastern region, the area affected by drought decreases in the past four decades (1971–2010) whereas the area affected by drought increases for the dry part, western and northwestern region. Climate simulation models showed the region likely to experience more frequent and severe droughts in the future (Sun et al., 2019). Together with rising temperature, the so-called hot drought would have more negative impact on crop production as the more than 80% of the cropland in the Loess Plateau is rainfed farmland.

The adaptive strategies employed by local farmers are influenced by a variety of factors, including their perceptions of climate change, its impact on crop yields, natural disasters, planting and harvesting schedules, and climate-induced pests and diseases. Additionally, demographic and social factors also play a role. Research indicates that while most farmers in the region acknowledge the negative effects of climate change on agriculture, they tend to adopt low-investment, low-risk, and quick-response measures, such as weather forecasting and adjusted agricultural practices. However, they are less likely to consider changes in crop varieties or land use structures (Shi et al., 2019). To address this, it is crucial to enhance farmers' awareness of climate change, promote government-led initiatives and subsidies for adaptive measures, and improve education levels.

13.4 Lower Reaches

The lower reach of the Yellow River spans 786 km, from Zhengzhou in Henan to the river's mouth at the Bohai Sea, encompassing the river delta and surrounding plain area. Research indicates that temperature has been steadily rising in the area, while precipitation remains relatively stable, despite substantial year-to-year fluctuations (Zhang et al., 2015a, 2015b; Mo et al., 2017; Xiao et al., 2020). Extreme weather events such as floods, droughts, and hailstorms are occurring more frequently, presenting challenges to agriculture. To adapt to climate changes, promoting "climate-smart" agriculture is helpful. This includes practices such as delayed seeding of winter wheat, tillage conversion, and water-saving irrigation methods.

Sea level rises could have a significant impact on the river delta area. The Yellow River Delta covers an area of 5,400 km^2, with its elevation gradually decreasing from the southern, older delta to the newly developed northern region, ranging from less than 1 meter to 13 meters above sea level. Satellite altimetry data shows that global sea levels have been rising at a rate of 3 mm per year since 1993. If this rate continues, sea levels will be 65 cm higher by the end of the century (Nerem et al.,

2018). The most recent IPCC report (AR6) projected that under the worst-case scenario, where global temperatures rise by more than 4 °C, sea levels are expected to increase by 0.6 to 1.1 meters by 2100 (IPCC, 2019). The rising sea levels will likely have a more significant impact on the northern part of the delta due to lower elevations, causing submersion in some areas. As sea levels continue to rise, coastal wave erosion is expected to intensify. Climate change not only affects the above-sea-level portions of the delta but also impacts the submerged areas. Jiang et al. (2017) studied the impact of climate change on the expansion of the submerged Yellow River Delta from 1976 to 2005, finding that the rapid increase in air temperature within the Yellow River basin contributed significantly to the transition from rapid accumulation (1976–1985) to moderate accumulation (1986–1995), and eventually to slow accumulation (1996–2005). With further temperature increases in the future, the submerged areas of the delta may enter a destructive phase. The rising sea levels will also lead to higher groundwater levels, and the intrusion of seawater into the groundwater will exacerbate soil salinization in the Yellow River delta wetland, particularly along the coastal areas (Xia et al., 2021; An et al., 2022). However, Guo et al. (2023) contended that over the past 40 years, human activities have had a more significant impact on the salinization of the delta than climate change, though these factors are not mutually exclusive. Salinization can harm wetland vegetation and increase greenhouse gas emissions from the wetland (Chen et al., 2018). Climate change in the middle reaches also impacts the development of the delta. Xu (2008) showed that since 1970, rising temperatures have weakened the East Asian Monsoon, leading to a decrease in precipitation. This has partly contributed to the decrease in sediment erosion from the Loess Plateau region, slowing the accretion rate of the delta.

13.5 Conclusions

The impact of climate change varies significantly across different regions of the Yellow River basin. In the river's source region, the high altitude means that rising temperatures could lead to permafrost thawing, which would affect carbon storage in the soil. The melting of snow and glaciers on the Tibetan Plateau will significantly influence flow discharge and its temporal variations in the upper streams. The middle reaches of the Yellow River, characterized by semi-arid and arid climates, will likely experience rising temperatures, more extreme storms, and droughts, which can have an impact on agriculture. Climate change may also affect soil erosion and water resources, although predictions of precipitation changes are inconsistent among different studies. Soil loss, driven by changes in rainfall and vegetation, involves a complex interplay of factors. As economic development continues, the demand for water will increase, further straining supply and availability. Additionally, water availability affects vegetation growth and reforestation efforts, and the succession of local ecosystems. The lower reaches of the Yellow River, especially the deltaic area, will be impacted by sea level rise due to global warming, which may exacerbate coastal erosion, affect wetland ecosystems, and increase soil salinization. It is essential to raise public awareness and improve education to better adapt to future climate changes.

References

An, X., Jin, W., Zhang, H., Liu, Y., & Zhang, M. (2022). Analysis of long-term wetland variations in China using land use/land cover dataset derived from Landsat images. *Ecological Indicators, 145*, 109689.

Chen, B., Zhang, X., Tao, J., Wu, J., Wang, J., Shi, P., Zhang, Y., & Yu, C. (2014). The impact of climate change and anthropogenic activities on alpine grassland over the Qinghai-Tibet plateau. *Agricultural and Forest Meteorology, 189*, 11–18.

Chen, Q., Guo, B., Zhao, C., & Xing, B. (2018). Characteristics of CH4 and CO2 emissions and influence of water and salinity in the Yellow River delta wetland, China. *Environmental Pollution, 239*, 289–299.

Cuo, L., Zhang, Y., Gao, Y., Hao, Z., & Cairang, L. (2013). The impacts of climate change and land cover/use transition on the hydrology in the upper Yellow River Basin, China. *Journal of Hydrology, 502*, 37–52.

Dong, S., Shang, Z., Gao, J., & Boone, R. B. (2020). Enhancing sustainability of grassland ecosystems through ecological restoration and grazing management in an era of climate change on Qinghai-Tibetan plateau. *Agriculture, Ecosystems & Environment, 287*, 106684.

Fang, Y., & Wei, Y. (2013). Climate change adaptation on the Qinghai–Tibetan Plateau: The importance of solar energy utilization for rural household. *Renewable and Sustainable Energy Reviews, 18*, 508–518.

Fang, K., Gou, X., Chen, F., Peng, J., D'Arrigo, R., Wright, W., & Li, M. H. (2009). Response of regional tree-line forests to climate change: Evidence from the northeastern Tibetan plateau. *Trees, 23*, 1321–1329.

Gao, G., Fu, B., Wang, S., Liang, W., & Jiang, X. (2016). Determining the hydrological responses to climate variability and land use/cover change in the loess plateau with the Budyko framework. *Science of the Total Environment, 557*, 331–342.

Gao, X., Zhao, Q., Zhao, X., Wu, P., Pan, W., Gao, X., & Sun, M. (2017). Temporal and spatial evolution of the standardized precipitation evapotranspiration index (SPEI) in the loess plateau under climate change from 2001 to 2050. *Science of the Total Environment, 595*, 191–200.

Guo, B., Liu, Y., Fan, J., Lu, M., Zang, W., Liu, C., Wang, B., Huang, X., Lai, J., & Wu, H. (2023). The salinization process and its response to the combined processes of climate change–human activity in the Yellow River Delta between 1984 and 2022. *Catena, 231*, 107301.

Han, Z., Huang, Q., Huang, S., Leng, G., Bai, Q., Liang, H., Wang, L., Zhao, J., & Fang, W. (2021). Spatial-temporal dynamics of agricultural drought in the loess plateau under a changing environment: Characteristics and potential influencing factors. *Agricultural Water Management, 244*, 106540.

Harris, R. B. (2010). Rangeland degradation on the Qinghai-Tibetan plateau: A review of the evidence of its magnitude and causes. *Journal of Arid Environments, 74*(1), 1–12.

He, X., Yan, J., Yang, L. E., Wu, Y., & Zhou, H. (2022). Climate change adaptation of smallholders on the Tibetan plateau under government interventions. *Journal of Cleaner Production, 381*, 135171.

Hu, Y., Maskey, S., & Uhlenbrook, S. (2012). Trends in temperature and rainfall extremes in the Yellow River source region, China. *Climatic change, 110*(1–2), 403–429.

IPCC. (2019). Summary for policymakers. In *IPCC special report on the Ocean and cryosphere in a changing climate*. IPCC.

Jiang, C., Pan, S., & Chen, S. (2017). Recent morphological changes of the Yellow River (Huanghe) submerged delta: Causes and environmental implications. *Geomorphology, 293*, 93–107.

Kong, D., Miao, C., Wu, J., Zheng, H., & Wu, S. (2020). Time lag of vegetation growth on the loess plateau in response to climate factors: Estimation, distribution, and influence. *Science of the Total Environment, 744*, 140726.

Li, Z., Liu, W. Z., Zhang, X. C., & Zheng, F. L. (2009). Impacts of land use change and climate variability on hydrology in an agricultural catchment on the loess plateau of China. *Journal of Hydrology, 377*(1–2), 35–42.

Li, Z., Liu, W. Z., Zhang, X. C., & Zheng, F. L. (2011). Assessing the site-specific impacts of climate change on hydrology, soil erosion and crop yields in the loess plateau of China. *Climatic Change, 105*(1), 223–242.

Li, L., Shen, H., Dai, S., Xiao, J., & Shi, X. (2012). Response of runoff to climate change and its future tendency in the source region of Yellow River. *Journal of Geographical Sciences, 22*, 431–440.

Li, G., Sun, S., Han, J., Yan, J., Liu, W., Wei, Y., Lu, N., & Sun, Y. (2019). Impacts of Chinese grain for green program and climate change on vegetation in the loess plateau during 1982–2015. *Science of the Total Environment, 660*, 177–187.

Liang, E., Wang, Y., Eckstein, D., & Luo, T. (2011). Little change in the fir tree-line position on the southeastern Tibetan plateau after 200 years of warming. *New Phytologist, 190*(3), 760–769.

Liu, X. D., & Chen, B. (2000). Climate warming in the Tibetan plateau during recent decades. *International Journal of Climatology, 20*, 1729–1742.

Liu, Y., Fan, J., Harris, W., Shao, Q., Zhou, Y., Wang, N., & Li, Y. (2013). Effects of plateau pika (Ochotona curzoniae) on net ecosystem carbon exchange of grassland in the three Rivers headwaters region, Qinghai-Tibet, China. *Plant and Soil, 366*, 491–504.

Liu, Y., Liu, L., Li, L., Li, H., Xu, H., Yang, J., Tao, S., & Zhu, B. (2023). Changes in runoff in the source region of the Yellow River Basin based on CMIP6 data under the goal of carbon neutrality. *Water, 15*(13), 2457.

Lu, W., Wang, W., Shao, Q., Yu, Z., Hao, Z., Xing, W., Yong, B., & Li, J. (2018). Hydrological projections of future climate change over the source region of Yellow River and Yangtze River in the Tibetan plateau: A comprehensive assessment by coupling RegCM4 and VIC model. *Hydrological Processes, 32*(13), 2096–2117.

Ming, J., Cachier, H., Xiao, C., Qin, D., Kang, S., Hou, S., & Xu, J. (2008). Black carbon record based on a shallow Himalayan ice core and its climatic implications. *Atmospheric Chemistry and Physics, 8*(5), 1343–1352.

Mo, X. G., Hu, S., Lin, Z. H., Liu, S. X., & Xia, J. (2017). Impacts of climate change on agricultural water resources and adaptation on the North China plain. *Advances in Climate Change Research, 8*(2), 93–98.

Nerem, R. S., Beckley, B. D., Fasullo, J. T., Hamlington, B. D., Masters, D., & Mitchum, G. T. (2018). Climate-change–driven accelerated sea-level rise detected in the altimeter era. *Proceedings of the National Academy of Sciences, 115*(9), 2022–2025.

Qin, Y., Yang, D., Gao, B., Wang, T., Chen, J., Chen, Y., Wang, Y., & Zheng, G. (2017). Impacts of climate warming on the frozen ground and eco-hydrology in the Yellow River source region, China. *Science of the Total Environment, 605*, 830–841.

Shi, X., Sun, L., Chen, X., & Wang, L. (2019). Farmers' perceived efficacy of adaptive behaviors to climate change in the loess plateau, China. *Science of the Total Environment, 697*, 134217.

Shi, S., Yu, J., Wang, F., Wang, P., Zhang, Y., & Jin, K. (2021). Quantitative contributions of climate change and human activities to vegetation changes over multiple time scales on the loess plateau. *Science of the Total Environment, 755*, 142419.

Sun, Q., Miao, C., Duan, Q., & Wang, Y. (2015). Temperature and precipitation changes over the loess plateau between 1961 and 2011, based on high-density gauge observations. *Global and Planetary Change, 132*, 1–10.

Sun, W., Mu, X., Song, X., Wu, D., Cheng, A., & Qiu, B. (2016). Changes in extreme temperature and precipitation events in the loess plateau (China) during 1960–2013 under global warming. *Atmospheric Research, 168*, 33–48.

Sun, C. X., Huang, G. H., Fan, Y., Zhou, X., Lu, C., & Wang, X. Q. (2019). Drought occurring with hot extremes: Changes under future climate change on loess plateau, China. *Earth's Future, 7*(6), 587–604.

Sun, Z., Liu, Y., Zhang, J., Chen, H., Shu, Z., Chen, X., Jin, J., Guan, T., Liu, C., He, R., & Wang, G. (2022). Projecting future precipitation in the Yellow River basin based on CMIP6 models. *Journal of Applied Meteorology and Climatology, 61*(10), 1399–1417.

Wang, Q. X., Fan, X. H., Qin, Z. D., & Wang, M. B. (2012). Change trends of temperature and precipitation in the loess plateau region of China, 1961–2010. *Global and Planetary Change, 92*, 138–147.

Wang, W., Zhao, X., Cao, J., Li, H., & Zhang, Q. (2020). Barriers and requirements to climate change adaptation of mountainous rural communities in developing countries: The case of the eastern Qinghai-Tibetan plateau of China. *Land Use Policy, 95*, 104354.

Wu, J., Miao, C., Zheng, H., Duan, Q., Lei, X., & Li, H. (2018). Meteorological and hydrological drought on the loess plateau, China: Evolutionary characteristics, impact, and propagation. *Journal of Geophysical Research: Atmospheres, 123*(20), 11–569.

Xia, C., Liu, G., Xia, H., Jiang, F., & Meng, Y. (2021). Influence of saline intrusion on the wetland ecosystem revealed by isotopic and hydrochemical indicators in the Yellow River Delta, China. *Ecological Indicators, 133*, 108422.

Xiao, D., Li Liu, D., Wang, B., Feng, P., Bai, H., & Tang, J. (2020). Climate change impact on yields and water use of wheat and maize in the North China plain under future climate change scenarios. *Agricultural Water Management, 238*, 106238.

Xu, J. (2008). Response of land accretion of the Yellow River delta to global climate change and human activity. *Quaternary International, 186*(1), 4–11.

Yang,K., Wu,H., Qin,J., Lin,C., Tang,W., Chen,Y., 2014. Recent climate changes over the Tibetan Plateau and their impacts on energy and water cycle: A review.

You, Q., Fraedrich, K., Sielmann, F., Min, J., Kang, S., Ji, Z., Zhu, X., & Ren, G. (2014). Present and projected degree days in China from observation, reanalysis and simulations. *Climate Dynamics, 43*, 1449–1462.

Zhang, X. C., & Liu, W. Z. (2005). Simulating potential response of hydrology, soil erosion, and crop productivity to climate change in Changwu tableland region on the loess plateau of China. *Agricultural and Forest Meteorology, 131*(3–4), 127–142.

Zhang, B., Wu, P., Zhao, X., Wang, Y., Wang, J., & Shi, Y. (2012). Drought variation trends in different subregions of the Chinese loess plateau over the past four decades. *Agricultural Water Management, 115*, 167–177.

Zhang, Y., Gao, Q., Dong, S., Liu, S., Wang, X., Su, X., Li, Y., Tang, L., Wu, X., & Zhao, H. (2015a). Effects of grazing and climate warming on plant diversity, productivity and living state in the alpine rangelands and cultivated grasslands of the Qinghai-Tibetan plateau. *The Rangeland Journal, 37*(1), 57–65.

Zhang, H. L., Zhao, X., Yin, X. G., Liu, S. L., Xue, J. F., Wang, M., Pu, C., Lal, R., & Chen, F. (2015b). Challenges and adaptations of farming to climate change in the North China plain. *Climatic Change, 129*, 213–224.

Zhang, T., Li, D., & Lu, X. (2022). Response of runoff components to climate change in the source-region of the Yellow River on the Tibetan plateau. *Hydrological Processes, 36*(6), e14633.

Zhao, A., Zhang, A., Lu, C., Wang, D., Wang, H., & Liu, H. (2017). Spatiotemporal variation of vegetation coverage before and after implementation of grain for green program in loess plateau, China. *Ecological Engineering, 104*, 13–22.

Zhou, D., & Huang, R. (2012). Response of water budget to recent climatic changes in the source region of the Yellow River. *Chinese Science Bulletin, 57*, 2155–2162.

Chapter 14
Syntheses and Conclusions

The Yellow River basin exhibits highly diverse physical conditions from its source to its outlet, which can be divided into three distinct regions. The source region is marked by high elevations, an alpine climate, and predominantly grassland and alpine meadow ecosystems. The middle reaches are mainly comprised of the Loess Plateau, characterized by a semi-arid climate and severe erosion. The lower reaches of the Yellow River are dominated by flat floodplains and a deltaic landscape. Correspondingly, there are different environmental problems facing those regions. The source region is threatened by grassland and wetland degradation, desertification, permafrost layer thawing, and biodiversity loss. The middle reaches suffer severe water and wind erosion, vegetation destruction, water resources shortage and water quality degradation; the lower reaches confront flood risk, groundwater dwindling, and salinization; The Yellow River delta faces issues such as wetland destruction and degradation, along with rising sea levels.

A cascading effect exists throughout the basin, from its source to its mouth. Water discharge in the source region, driven by precipitation and snowmelt, influences conditions downstream. Severe erosion in the middle reaches causes sediment depositions in the lower channel and reservoirs, affecting floodwater flow and contributing to the formation of the delta. Pollutants from non-point sources, such as agricultural runoff, as well as point sources like industrial activities and domestic water use, enter the river network, resulting in water quality deterioration downstream.

Although numerous studies have been conducted on environmental issues in the basin, they have limitations and yield controversial results, highlighting the continued need for future research. In the source area, the uplift of the Tibetan Plateau has had a profound effect on the climate and environment. The stages of uplift and their influence on the progression of paleoclimate remain a subject of debate and need to be further studied. There is considerable uncertainty about how the Tibetan Plateau influences the East Asian monsoon climate. Many paleoclimate reconstruction studies rely on samples from a limited number of sites and then extrapolate the findings

T. Zhu, *Watershed Environmental Degradation and Sustainable Development*, https://doi.org/10.1007/978-3-032-02448-0_14

to the entire region, which can lead to significant errors. There is very limited integration of paleoclimates into future climate changes predictions. Further research is also needed to understand the impact of climate change on permafrost, including its severity, extent, and the greenhouse gas emissions resulting from thawing. The relative effects of human activities and climate change on grassland degradation, wetland deterioration, and biodiversity remain subjects of ongoing debate. In the middle reaches, paleoclimate reconstruction mainly relies on loess profiles. However, the limitations lie in the very few loess profiles in those studies and the proxies used to study the paleoclimate. The identification of the region's natural vegetation prior to the large-scale human activities is largely based on historical documents, but these sources are often site-specific, leading to significant uncertainty. The effect of climate change on the environment over the past few decades is primarily assessed using an extensive network of weather stations, but the research results are inconsistent and even controversial among different studies. There have been limited studies to predict future climate changes and the results are widely varied, further reducing confidence in such predictions. Soil erosion is the primary issue in the region. While numerous studies have been conducted on sheet and rill erosion, less attention has been placed on other forms of erosion, such as gully erosion, mass movement, tunnel erosion, and wind erosion. Studies on soil erosion during extreme events are limited. The sediment delivery ratio is another important topic that warrants more attention in future studies. Water resources in the region are generally very limited and long-term studies are necessary to assess the impact of afforestation on water resources. Further research is also needed on the reclamation of mining areas and the effects of coal mining on water quality, especially ground water, and soil erosion. In the lower reaches, additional studies are required on channel sedimentation and flood risks, and the impact resulting from sea-level rise. A common issue with current studies across the basin is that many existing field studies are short-term, often yielding results that are limited to specific areas and timeframes. As a result, these studies may produce differing and even controversial outcomes, making them questionable for application to other regions or over extended time periods. Therefore, it is important to conduct long-term studies and establish more research stations throughout the basin. Interdisciplinary research that integrates the natural sciences, social sciences, and economics is essential, as environmental planning based solely on natural sciences often faces limitations due to social and economic factors. Combining these fields can result in more effective and practical solutions to environmental challenges.

Significant efforts have been made in the basin to restore the environment over recent decades. However, concerns and debates persist. First, large-scale tree planting consumes significant amounts of water in areas where water resources are already scarce. Some argue that its long-term sustainability is questionable and suggest that regional variations in natural conditions, particularly precipitation, should be considered in revegetation efforts. Second, ecological restoration must be aligned with food security, economic development, and improving people's living standards. While government funding and subsidies are essential to kickstart these initiatives, long-term success ultimately depends on the active involvement of local farmers to maintain and sustain their efforts. Third, climate change presents challenges and uncertainties for watershed management. While there is general agreement that temperature will keep rising, there is no consensus on how precipitation patterns will shift, a crucial factor for environmental sustainability, agricultural production, water resource management, and economic development. Adequate preparations to address these challenges have not yet been made. Fourth, certain areas facing severe erosion, such as gully regions, wind-water intersection zones, the Pisha sandy area, and coal mining regions, will require ongoing control efforts in the future.

To achieve sustainable development in the region, environmental remediation must be closely linked with economic growth and the well-being of local communities. Sustainable agriculture should aim to produce enough food for local populations while preserving or enhancing environmental conditions. This can be achieved through land-use changes, such as developing high-yield agricultural lands like check-dams and terraced lands. Vegetation recovery should combine natural regrowth with artificial planting efforts. In gully areas, natural recovery through grazing prohibition may be more cost-effective. For artificial planting, promoting biodiversity by using a variety of species is important. Water resources are among the most limiting factors for regional development in the basin. In the long term, the region will require water transfers from other watersheds to address the permanent water shortage issue. In the short term, effective water resource management is essential, necessitating coordination between upstream and downstream areas to balance the needs of agriculture, industry, households, and ecosystems. In coal mining areas, stricter policies must be implemented to prevent further environmental degradation. In conclusion, tackling the region's environmental challenges demands a holistic and integrated approach. The progress towards sustainability should consider local economic conditions, employment opportunities, farmers' household incomes, and population trends. Enhancing environmental awareness and promoting sustainability through education are key to encouraging local communities to embrace sustainable practices.

Index

T. Zhu, *Watershed Environmental Degradation and Sustainable Development*, https://doi.org/10.1007/978-3-032-02448-0